FLIGHT OF THE LAVI

FLIGHT OF THE LAVI

Inside a U.S.–Israeli Crisis

Dov S. Zakheim

BRASSEY'S
Washington • London

Library of Congress Cataloging-in-Publication Data

Zakheim, Dov S.
    Flight of the Lavi: inside a U.S.–Israeli crisis/Dov S. Zakheim.
      p.   cm.
    Includes index.
    ISBN 1-57488-065-9
    1. Lavi (Jet fighter plane) 2. Military assistance, American—Israel.
3. United States—Foreign relations—Israel. 4. Israel—Foreign relations—United States. I. Title.
UG1242.Z35   1996
355'.0325694—dc20   96-1284

10 9 8 7 6 5 4 3 2 1

Printed in the United States of America

For my Parents

Contents

Foreword by Caspar Weinberger   ix
Preface   xiii

1   Israel's New Fighter   3

2   Something About This Program Stinks   15

3   Getting Down to Business   32

4   Meetings, Meetings—in Washington, in Israel   48

5   Preparing the Lavi Report   65

6   Israel Responds to the Lavi Report   86

7   Rabin Blesses a Second Lavi Report   95

8   The Lavi Gets Personal   109

9   Two Sides of Dual Loyalty:
    Cap Weinberger and the Pollard Affair   121

10 Run-up to Another Study Visit   126

11 The Debate Heats Up   149

12 Enter the GAO   159

13 Rollout   164

14 The Navy Study Is Completed   185

15 We Finish the Alternatives Study   202

16 Exiting DoD   225

17 The Lavi Is Grounded   238

18 Aftermath   253

Notes   259
Index   268

Foreword

F*light of the Lavi* is a very important book because it takes us inside the highest governmental levels of two close allies, the United States and Israel, as they dealt with a serious disagreement. In Dov Zakheim's intriguing narrative of his personal involvement, the reader is introduced to the key players and to how high-level decisions are made and influenced.

Beginning in 1983, the Israeli embassy and American friends of Israel worked with key congressional staff members and politicians to introduce some unusual modifications to the U.S. military assistance program. The proposed legislation supported the sharing of specific advanced technologies and identified large amounts of U.S. funding to be allocated to the Israeli development of a new advanced fighter aircraft, the Lavi. I was told of many misgivings of my staff to this legislation. I agreed with Fred Iklé, our undersecretary for policy, that we needed an objective U.S. assessment of the proposed Israeli fighter aircraft program. Fred recommended Dr. Dov Zakheim (at that time our assistant undersecretary for policy/resources) to lead this assessment. I immediately agreed. Dov seemed ideal. He was a former congressional staffer with excellent analytical abilities and good contacts on the Hill and in the media. He had earned a reputation in the Defense Department for effective action and negotiating skills. An Orthodox Jew, he spoke Hebrew fluently. At the time I was unaware that Dov's father had grown up in a small European town with several of the political leaders of Israel. I also could not foresee the trauma this difficult assignment would bring to him and his family.

This book tells the rest of that story. Dov very quickly found and documented that Israel and its U.S. friends already had effectively mobilized our Congress and the Department of State for support. He then began a two-year odyssey of investigation, negotiation, and persuasion. In the process, he faced strong enemies and found sturdy allies at the top of the Israeli political, defense, and industrial sectors.

Dov found the Lavi to be a project requiring the investment of billions of U.S. taxpayers' dollars. Moreover, he learned the Israelis were planning to sell the Lavi—and its advanced American-designed weapon systems—to other countries in direct competition with the very U.S. aerospace companies that would provide their proprietary technology.

This book describes Dov's findings that the objectives of the fighter development went well beyond export revenue and included Israeli national pride, stimulus to the high-technology sector of Israel's economy, and making the country more self-sufficient in its defense. The driving force, however, for the new plane came from Moshe Arens, a former defense minister, who held a doctorate in aeronautical engineering from the Massachusetts Institute of Technology. He was strongly supported by Israeli labor unions and the industries involved in the Lavi development.

Zakheim describes an intriguing series of U.S. and Israeli studies, and the difficulties of obtaining information necessary to answer his inquiries. Nevertheless, Zakheim's expert team, after a full and careful analysis, concluded that the Lavi would be far over budget and would cost significantly more than if the Israelis continued their earlier policy of buying additional advanced U.S. fighters to add to those already in their inventory.

The fight then began in earnest. Zakheim reports his amazement that the Israeli development team felt sure it could persuade the United States to pay for all cost overruns on the proposed new plane. Zakheim is forced to argue against the Lavi on two fronts: at home in the United States and in Israel's political and defense systems. His story of personal meetings with members of the Israeli government, business, and media are fascinating. Particularly revealing are his private meetings in the homes and offices of three Prime Ministers, Yitzchak Rabin, Yitzchak Shamir, and Shimon Peres.

The political battle ebbed and flowed as Zakheim tried to bolster allies inside the Israeli government and armed forces by reinforcing their concerns about how the growing costs of the Lavi would starve the army, the navy, and civilian social programs. I find particular poignancy in his book because Dov never told us how this "assignment" was deeply affecting him emotionally in both positive and negative ways. His narrative shares with us how his required trips to Israel permitted intense personal communion with the land of his religion and reinforced his personal spirit. But he also found that those Israelis who saw him as "a traitor to the family" would adopt the biblical adage of "an eye for an eye and a tooth for a tooth." He reports how he and his family ran into severe criticism bordering on harassment in the media of both countries, at synagogue, at meetings of Jewish organizations, and even from his children's classmates in elementary school.

The fate of the Lavi was finally decided in a suspenseful, contentious, extremely close vote of the Israeli cabinet. Readers of this book can enjoy the extraordinary insider's perspective it provides about how Israel and the United States influence each other's governments and decisions. They can also be grateful, as I am, that we had Dov Zakheim as an American official who did what was best for the United States—and for Israel.

Flight of the Lavi is a captivating story of courage, loyalty, passionate beliefs, and fateful decisions. I was there and I know it to be true. All those who want to understand better how political systems *really* work should read this book.

CASPAR WEINBERGER
Washington, D.C.

Preface

On a mild Tel Aviv November night in 1995, at a huge rally in support of the Middle East peace process, a lone gunman, an Orthodox Jew, gunned down the prime minister of Israel, Yitzchak Rabin. I was in Tunisia at the time, having been in Israel just a few days earlier. A Tunisian friend, a Muslim, tried to console me: "I know you're alone here in Tunisia, and that you're hurting. I am hurting with you." I was overwhelmed by his remark, both because it was so reflective of the peace that was beginning to envelop the troubled Middle East and because it evoked my personal memories of Israel's fallen leader.

I had come to know Yitzchak Rabin quite well, primarily professionally, but to a not small extent personally as well. I had worked with him, at times seemingly at cross purposes, but in the end very much in concert, to defeat the Lavi, an Israeli aircraft development program that was being funded by hundreds of millions of American taxpayer dollars, when cheaper American-built alternatives were readily available.

The Lavi was more than just a very costly weapons project, however. To its supporters, it was a manifestation of Israel's growing technological prowess and its ability to develop advanced weapons systems independently, without reliance on any other state. Paradoxically, it was also in many ways the culmination of a pattern of increasing American support for Israel that had begun in the after-math of the June 1967 Six-Day War and had accelerated after the signing of the Camp David Accords and Israel's peace with Egypt in 1979. The United States had become the major guarantor of Israel's security and had defined security as maintaining Israel's "qualitative edge" over its Arab adversaries. The Lavi, which drew upon Ameri-can high technology and American research dollars, was seen as the latest expression of that edge.

As the cost of weapons and their technological complexity had grown, so too had the intricate web of security commitments and relationships that marked American policy toward Israel. Israel was not an American ally in the formal, legal sense of the term. The rela-tionship with America also suffered from frequent irritants such as Israel's bombing of Iraq's Osirak reactor in 1981, its invasion of Lebanon and confrontation with American forces there in 1982, and frequent American arms sales to Arab states. Still, the personal ties between the leading elites of both countries were remarkable. The fre-quency with which leading politicians from both states visited each other's capitals was virtually unmatched in relationships between the United States and any of its other allies. So too was the general degree to which each country figured in the workings of the other's political system, extending even to the private lives of officials in both.

Perhaps more than any other project, the struggle over the Lavi illustrated the nature and unique intimacy of those ties. It also under-scored the problems they could engender as a result of the fundamen-tally different notions of cost and benefit, security and national pride, that governed perceptions in Washington and New York, Tel Aviv and Jerusalem. Conceived in Israel but implemented in the United States by a coalition of Israeli and American politicians and staff-level operatives, the Lavi became entangled in Israeli partisan politics, in American pressure-group politics, and even in the frictions between the Reagan administration's departments of State and Defense.

I was thrown into this maelstrom of pressures and counter-pressures several years after the project had been started. What the Israelis have come to call *parashat ha-lavi*, the Lavi episode, was for me far more than just another aspect of my duties as a fourth-tier Pentagon official. I found myself in direct conflict with powerful lead-ing Israeli politicians with whom I was personally linked, albeit indi-rectly. In fact, the Lavi's most vociferous Israeli proponent, Moshe Arens, himself a former defense minister and, during my involvement with the project, a minister without portfolio, was the son of one of my father's former business clients. My father had known Moshe Arens since he was a boy. In Arens's eyes, my opposition to the Lavi was an unforgivable sin. He had once told Yitzchak Rabin, his successor at the defense ministry (and Rabin later told me), "Zakheim is a traitor to the family."

The language Arens used was ugly; Rabin was himself shot down by people who also freely used the same epithet. But it was clear that my involvement in the fight to stop the plane did have a certain "man-bites-dog" quality about it. The media, ever attuned to every nuance of American-Israeli relations, quickly seized on the fact that it was an American Jew who was taking on Israeli officialdom. And not just an American Jew. I am a practicing Orthodox Jew, born in Brooklyn, and I speak Hebrew fluently, albeit with a horrible American accent. I was educated in Jewish schools and attended a Zionist summer camp. My father has been active in Zionist affairs all his life: he numbered Menachem Begin among his friends and grew up in the same small Polish town as Yitzchak Shamir. It was almost as if I fit the stereotype of the West Bank settler. Small wonder, then, that I found my life turned upside down as I was assigned to investigate and then lead the opposition to the production of Israel's prized plane, which threatened to demand billions more U.S. taxpayer dollars before it was complete.

It may be that the Lavi episode was unique in the annals of American-Israeli relations and never to be repeated under any circumstances. I doubt it. The intimate connections between Israel and the United States are such that even lower-level officials can continue to determine the outcome of major decisions in each other's countries. To ensure that citizens of both countries understand how these complex and often difficult and painful relations are conducted, I have written this insider's view of a major crisis between Israel and the United States.

Israel is today building a major antiballistic missile system, the Arrow, funded primarily with hundreds of millions of American taxpayer dollars. The program has been bolstered by support from many of the same political elements, in some cases the very same people, who put the Lavi on the Israeli-American agenda. The Arrow has thus far avoided the controversy that surrounded the Lavi, partly because there is no real American alternative to it. If, as I hope, the U.S.–Israeli relationship endures, there may be other jointly developed, American-funded projects that will come to the fore. With luck, they will avoid the pitfalls that ultimately grounded the Lavi.

Luck will not be enough, however. What will also be required, in large doses, will be knowledgeable and courageous leadership in both the United States and in Israel. Leadership such as that which Defense Minister Yitzchak Rabin provided when he patiently and cautiously, yet with a shrewd understanding of both American and Israeli political gamesmanship, reversed Israel's nearly decade-long decision to build the Lavi. Leadership such as that which Rabin, the former chief of staff and war hero, afforded his country and the world when, as prime minister, he extended his hand in friendship to his former Arab adversaries, and then sacrificed his life on the battleground of peace.

Thus, I solemnly dedicate this volume to Yitzchak Rabin's blessed memory, to the unique relationship between his country and mine, and to my family, which stood behind me during a very trying time when that relationship was sorely tested.

Acknowledgments

I am fortunate to have benefitted from the encouragement of several people as I waded through the mass of paper, the press reports, and my own recollections to produce this volume. My wife, Deborah, and my sons were stalwart supporters. My eldest son, Chaim, read many of the early draft chapters with enthusiasm. My friend, and deputy both at the Pentagon and at System Planning Corporation, Dick Smull, read the entire manuscript. Frank Margiotta, a friend of nearly two decades and president of Brassey's, Inc., provided priceless advice; Carsten Fries and Don McKeon were a tremendous help, while the Brassey's editorial staff were rigorous, invaluable literary stylists. Finally, I wish to thank the Almighty, Who sustained me in what was both an adventure and an ordeal, "and was with me in the way which I went" (Genesis 35:3).

FLIGHT OF THE LAVI

1

Israel's New Fighter

I knew very little about the Lavi, other than that *lavi* meant "lion" in Hebrew, when I learned in early March 1985 that my boss, Under Secretary of Defense Fred Iklé, was planning a trip to Israel the following month. I hadn't been to Israel in nearly a decade, and at that time, as always in the past, solely as a tourist. I had minimal professional association with Israelis, the only exception having been my championing the Pentagon's purchase of Israeli remotely piloted vehicles in the aftermath of the Lebanon War.

My remoteness from Israeli politico-military affairs had been quite deliberate. I felt that it was virtually impossible to deal with Middle Eastern matters in a rational way: they were permeated by ideology, theology, or both. For me, the State of Israel was not a religion; at best it was a manifestation of religion. I did not even see myself as a Zionist per se, though Arab acquaintances of mine labeled me a Zionist simply because I maintained that the state had a right to exist. I felt, however, that the essence of Zionism was a belief that all Jews must return to Israel, and I could not subscribe to that belief. I have always believed that there will be a Diaspora until the coming of the Messiah, and I have no great insights as to when precisely that might occur.

Given my somewhat unconventional attitudes to Israel, I always considered it prudent to turn my research and professional interests

3

elsewhere. But I had a strong sentimental tie to the place and had very much enjoyed my visits there, which began just after the June 1967 War. So when I learned that Iklé was planning his trip, I asked if I could come along.

My relationship with Iklé was quite good, and I fully expected him to take me along. What I didn't realize was that there would be a condition attached to the trip. Apart from being an internationally recognized defense expert, Iklé is an extremely shrewd man; he usually conceals this shrewdness behind a poker-faced, somewhat acerbic Swiss demeanor. I had developed a small reputation in the Pentagon as something of a troubleshooter, and had just successfully revived—and was on the verge of completing—what had previously been a moribund negotiation with Canada on the upgrading of the joint U.S.–Canadian North Warning radar system. Iklé had in mind more trouble for me to shoot at; how much trouble was something that I believe he recognized from the start, but I understood only years later.

Iklé told me that I could come along with him, but that he had a project in mind for me. He told me that the Israelis were developing an airplane supposedly on their own, but with a rather large financial input from the United States. Iklé thought the plane was a waste of U.S. money and Israeli time. In his view the money, drawn from American grant assistance, which had become increasingly generous during the early 1980s, was better spent on other Israeli programs, as well as on the acquisition of an American fighter aircraft. Iklé wanted to stop the project, but was not sure how that might be done. Caspar Weinberger, the secretary of defense, had tried, and failed, to slow it down. Iklé wanted me to give the program one last look and, if I could do so, develop and implement a plan to terminate it. It was a very tall order.

Were Israel any other country—Sweden, for example, which was also developing a new aircraft—the matter would have been quite straightforward. Israel could go ahead on its own, and U.S. funding would be provided for other projects. But Israel is not any other country, of course. And the United States had been intimately involved with the project both technically and financially virtually since its inception.

The origins of the project dated back to the French embargo of Israel in the aftermath of the Six-Day War. (The embargo was still in effect in the mid-1980s.) The Israelis had concluded that they had to develop their own aerospace industry, that they could never again solely rely on "friends" from abroad. The most prominent exemplar of the new Israeli emphasis on indigenous military industrial development was Israel Aircraft Industries. IAI had been founded in the 1950s by an expatriate American and built aircraft, missiles, and lesser products; it serviced these systems and provided many of their components through its own subsidiaries.

By the late 1970s, IAI had built its own fighter plane, the Kfir, from a French Mirage design, which it had sold to the Israeli air force and which it was marketing worldwide. For its part, the Israeli air force began planning for a Kfir replacement, which it hoped would also serve as a replacement for its fleet of small A-4 attack aircraft, purchased from the United States in the 1970s. Given IAI's success with the Kfir, it wasn't surprising that the Israelis preferred to replace it with another indigenously built aircraft. The problem, however, was that such a plane would not be truly indigenous, any more than the Kfir or indeed other aircraft built outside the United States or the Soviet Union were likely to be. The Israelis recognized that they would have to look overseas for the plane's engines, as well as for other key components. Indeed, the more sophisticated the plane they wished to build, the more dependent they would be on foreign support.

To minimize their degree of dependence on foreign suppliers, the Israelis conceived of a relatively simple plane, termed the Aryeh, that would capitalize upon the technical advances that IAI was expected to have achieved in the 1980s but would nevertheless remain on the low end of the spectrum of sophistication associated with ground attack aircraft. It was in that spirit that Minister of Defense Ezer Weitzmann approached his American counterpart, Harold Brown, in April 1980, to obtain American support for the coproduction of General Electric F-404 engines in Israel.

Brown assented to Weitzmann's request, but with several critical technological, financial, and political reservations. He insisted that certain sensitive engine components would have to be produced in the United States. He refused to release any American foreign military sales (FMS) credits for purchases of Israeli goods or services for the aircraft. These credits were normally provided to Israel, as well as other foreign countries, for the purchase of American military equipment. Finally, Brown denied the release of any FMS credits for any portion of the aircraft that might then be reexported by the Israelis to third countries.

Weitzmann, nephew of Israel's first president (and since 1992 himself president of Israel), former air force commander, flamboyant hero of the Six-Day War, and hawk-turned-dove, agreed to these conditions without reservation. He recognized the need to assuage American fears about Israeli competition in sales to other countries and, in any event, did not have sufficient confidence in the ability of Israeli industry to produce a complex, multirole aircraft that could compete with the newest generation of American fighter and attack planes. Weitzmann hoped that Israel would develop a prototype plane by 1984 and would begin production by 1988.

The plane was renamed the Lavi, a more appropriate appellation relative to its mission. *Kfir* in Hebrew is a young lion; *aryeh*, on the other hand, is a mature king of beasts. A *lavi* is more mature than a *kfir*, but still not as developed as the fully maned variety—and the new plane, as Weitzmann conceived it, was indeed to be the junior partner in Israel's air force.

Israel had already begun a program to acquire more sophisticated F-15 and F-16 fighters from the United States. It also anticipated employing F-16s in an air-to-ground role as well, as the 1981 Israeli F-16 bombing of Iraq's Osirak reactor spectacularly demonstrated to the world. Weitzmann's concept, therefore, was for the modernization of Israel's air force with what military analysts termed a "high-low" mix: the top-of-the-line F-15s and F-16s would provide Israel with air superiority and deep strike capability; the yeoman's work of supporting the ground battle would be undertaken by the new plane.

Weitzmann resigned from the defense ministry in May 1980, in protest over the government's policies regarding West Bank autonomy and the expansion of settlements. His departure signaled a fundamental change in the nature of the aircraft. Weitzmann's successor, Ariel Sharon, was an armor officer who never claimed to be an expert on aircraft. Both he and Prime Minister Menachem Begin were therefore swayed by the arguments of Moshe Arens, a leading Likud Party figure and the chairman of the Knesset's foreign affairs and defense committee (and opponent of the Camp David Accords), that what Israel really required was an ultrasophisticated fighter/attack plane, at least on a par with the F-16, and probably superior to it.

Arens was not an uninterested party. He was a former senior official of Israel Aircraft Industries. Moreover, he was an aeronautical engineer by profession. He had earned his doctorate at MIT and had taught aeronautical engineering at the Technion, Israel's world-class technical institution of higher learning. Engineers have a strong tendency to believe that they can develop anything, and that anything they and their partners develop will *ipso facto* be more capable than anything the competition might develop, whoever the competition happens to be. Some whispered that Arens had another motive: his career at IAI had not been notably successful, and his detractors (of whom there were many in Israeli political circles) whispered that he was trying to "make it up" to IAI.

Whatever the reason, Arens's case prevailed, and Ariel Sharon approached the new American defense secretary, Caspar Weinberger, with a request for additional support for the project. Sharon wanted Weinberger to grant export licenses for the wing and vertical tail composite design produced by the Grumman Corporation of New York; servo-actuators, produced by the Moog Corporation, also of New York; the Lear Siegler flight control computer; and a variety of software tools. He also had a list of future license requests that he was expecting to forward early in 1983. These and other subsystems and design capabilities underscored the Lavi's utter dependence for its success on not only American dollars, but also American technological know-how.

The courtly Weinberger, whose Middle Eastern ties prior to entering office were exclusively with the Arabs and primarily with the Saudis, viscerally disliked the bumptious Israeli minister. He was to clash time and again with the former general; the word around the Pentagon was that no one had ever seen Cap lose his temper, except when he lost it arguing with Ariel Sharon. Not surprisingly, Sharon's request was denied.

There were other reasons for Weinberger's opposition. A number of American defense companies, most notably Northrop, vehemently opposed the creation of yet another competitor for their products. Northrop was hoping to market its own F-5G fighter (later renamed the F-20) overseas, as a replacement for its wildly successful F-5. The Lavi threatened to be the F-5G's primary foreign competitor, particularly in Latin America and possibly in East Asia as well.

Northrop was aided by the silence of key defense figures, particularly John Lehman, secretary of the navy, who otherwise were known to be solidly pro-Israel. But John had been a Northrop consultant prior to joining the Reagan administration in early 1981, and he (quite properly in my view) kept himself away from the Lavi debate.

At the same time as Sharon was pressing Weinberger for additional technical support for the project, Arens, now Israel's ambassador to the United States, began to marshal financial support for the project on Capitol Hill. The Hill has traditionally been Israel's best friend. Countless books, articles, and doctoral dissertations have analyzed just why it is that Israel is so successful at obtaining whatever it wishes from the Congress, often despite bitter opposition from the executive branch. Whatever the reasons, the reality cannot be denied, and Arens clearly exploited his advantage.

In February 1983, after a judicial panel asserted that he had been indirectly responsible for the massacre of Palestinians in the Lebanese refugee camps of Sabra and Shatila, an unrepentant Ariel Sharon resigned as defense minister. Moshe Arens replaced him. Not long before he returned to Israel, Arens hit upon the idea of the United States providing direct support for research and development on the Lavi *in Israel.* The notion was almost revolutionary. Heretofore, and as Harold Brown had decided, even in the case of sales to Israel, American security assistance took the form of grants or loans that were tied to the purchases of American equipment. Dollars were to be spent at home; if the buyer could not afford those dollars, the U.S. government, through its assistance programs, would pick up the tab. Arens wanted the dollars to go abroad, however. American industry and workers would see no immediate benefit from those dollars. Some critics argued that the American economy would see no benefit at any time. Arens was undeterred.

Arens worked especially closely with a congressman from east Texas, Charles Wilson. Unlike other politicians and their staffers who were sensitive to Jewish votes or financial support, Wilson would have done just as well politically without supporting Israel. In fact, at first glance, the partnership between the jovial Texan and the dry professor-turned-diplomat seemed a most unlikely one. But the two men did hit it off, and Wilson, convinced by the need to do everything possible to strengthen the Jewish state—which Arens interpreted for him to mean strengthening the Lavi program—crafted the plan for an amendment to the fiscal year 1984 foreign aid bill to allow Israel to spend foreign military sales credits on the Lavi in Israel.

Of course, Wilson had no idea just how much the Israelis needed. He therefore turned to the American Israel Public Affairs Committee, with whose leaders Arens also had excellent ties. AIPAC had long been recognized as one of the most effective lobbying organizations in Washington. Long known for its clout with legislative offices, which recognized that its endorsement could be critical to financial support and electoral success, AIPAC had recently expanded its staff to include military systems analysts, so as to bolster the "objective" case for support to Israel. It had also moved to new, impressive quarters just off Capitol Hill and had added lobbying of administration officials to its long-standing Hill lobbying mission.

While AIPAC was prepared to help Wilson derive an appropriate figure for his amendment, it had no clearer idea of how much was needed than he did. AIPAC therefore turned to Marvin Klemow, IAI's veteran Washington representative, who knew how to play the political game as well as anyone. I don't know whether Klemow or someone else actually proposed the figure that ultimately found its way into Wilson's amendment: $550 million for the Lavi, of which $300 million would be spent in Israel. Whoever derived the number had a good imagination, however. One Israeli embassy official later recalled, "The figure came right out of thin air."[1]

With the text of the amendment settled, Arens and Wilson solidified their powerful coalition in support of the proposal, whose only precedent was a much smaller 1977 grant of $30 million for development of the Merkava tank in Israel. While AIPAC remained a critical source of support, it was only one of the allies to which Arens turned. The Hill was stocked with staffers with whom Arens was extremely friendly. One of them, Jim Bond, proved especially helpful. Bond was Senator Bob Kasten's key staffer on the Foreign Operations Subcommittee of the Senate Appropriations Committee. With the Republicans in control of the Senate and Kasten chairing the subcommittee, Bond had considerable clout over the foreign aid budget. Kasten was a strong supporter of Israel and had many Jewish connections both in Wisconsin, his home state, and in Washington. He was very close to AIPAC, as was Bond. What AIPAC suggested, Bond could make happen; indeed, Bond had worked with Wilson in first crafting the House amendment.

In the end, Charlie Wilson's name did not appear on the amendment that created the offshore assistance program for the Lavi. The congressman claims no hard feelings over the fact that others attached their names to the legislation he had conceived.[2] That others were so eager to attach their names to the proposal and that Arens could fashion the sponsorship so that it was bipartisan testify to the political astuteness as well as determination of the Israeli ambassador-turned-defense-minister. An amendment to stop foreign offshore funding was roundly defeated. (The following year another such amendment was withdrawn, with its sponsor explaining, in a deliciously mixed metaphor, "I didn't want to be a goy kamikaze."[3] In the event, the Congress was guilty of overkill. As Klemow later put it, "We couldn't spend it all."[4]

Within two months of taking office, Arens had engineered another coup for the Lavi. Caspar Weinberger had viewed the program with suspicion from the moment it was briefed to him, and not only because it was Ariel Sharon who was its sponsor when he first addressed the issue. Thus when Weinberger blocked the export licenses required by American companies bidding for Lavi contracts, he threatened to slow the program's progress, if not undermine it entirely.

On April 13, 1983, Arens chaired a five-hour meeting in Jerusalem to discuss how to obtain the necessary licenses.[5] Among those attending the meeting were not only about twenty members of the Lavi project team but Danny Halperin, Israel's economics minister in Washington, a lean, wiry man with narrow eyes who had worked closely with Arens during his tenure as ambassador, and Marv Klemow, the portly, affable IAI representative. Like Marv, Halperin also was adept at the Washington game, probably more so than virtually any foreign diplomat. Both men had many friends in the press and on Capitol Hill; they got along well with each other, and had flown to Israel together from Washington. It was therefore a group of close comrades in arms that met in Arens's office that day.

Eight months earlier, Klemow had pressed the Israelis to attempt to convince the International Security Affairs (ISA) staff at the Department of Defense as well as the Israel desk at State to support the Lavi. Now, however, he recommended that Arens bypass Defense. He was later quoted as having told Arens, "Our strategy should be that the Pentagon doesn't exist. This is a political decision." Halperin agreed, and urged Arens to phone Secretary of State George Shultz to break the contracts logjam. Arens concurred. That evening he spoke to Shultz and requested his help. Within forty-eight hours the licenses were unblocked, and the Lavi program moved ahead. It was widely assumed that President Reagan personally intervened to break the impasse.[6]

Arens made one other decision in April 1983 that was critical to the Lavi's future. He brought David Ivri back into the ministry of defense (MoD). Ivri was a former air force commander who was serving as chairman of Israel Aircraft Industries, the Lavi prime contractor. Arens named him deputy chief of staff of the Israel Defense Forces. Ivri subsequently returned to IAI, but in mid-1986 returned once again to the MoD as its director general. He proved to be as tenacious as Arens in his support of the Lavi program. Like Arens, he clung to the fantasy that the Lavi was an Israeli national program that was essential to Israeli national honor.

In the late summer of that year, Menachem Begin resigned as prime minister of Israel as a result of Israel's increasingly bitter fruits of victory in Lebanon. He was replaced by Yitzchak Shamir, who, though foreign minister, was still very much an unknown quantity to the United States, in part because of his poor command of the English language. Arens was retained as Shamir's defense minister.

Arens and Shamir went back many years; he proved over time to be Shamir's closest adviser. My father knew them both, from his days in Lithuania as a leader of the Betar movement, to which all of them belonged. Shamir in fact grew up in my father's own home-town, Rozhenoi, a rather small place where, as in American small towns, everybody knows everybody else. He was a classmate of one of my aunts, and according to an Israeli newspaper report was in some way romantically involved with her. I never learned the source of that one—the Israeli press is notorious for its inventiveness—but my Dad emphatically denies that there was much between them. In any event, my aunt was killed by the Nazis during World War II.

Shamir's background was intelligence; he had been an agent of the Mossad. His best years were spent in France, as he wistfully recalled to me during a private evening I had at his home in 1986. Shamir was also a key leader of Lehi—the Hebrew acronym for three words that mean "fighters for Israel's freedom." Lehi was better known as the Stern Gang, a group of violent extremists who were despised by the mainstream Zionist organizations and their military arm, the Haganah. Many Lehi personnel moved into intelligence at the war's end.

Not surprisingly for a man with a lifetime's background in covert operations of one sort or another, Shamir was not one to trust many people. He had even met his wife in the resistance underground. But he trusted Arens, particularly on security matters, despite Arens's own very different background as a bookish Latvian-born but MIT-trained professor of aeronautical engineering who had emigrated to Israel only after first serving in the American army.

Arens was also a Betari. His father was one of my father's clients.

Together with a cousin, his father had started a plastics company. As a small boy I often heard my father talking about Arens and his partner, though I hadn't the foggiest idea who they were. I never met them, and only years later, as a defense official, did I learn of the connection between my father's clients and the defense minister of Israel.

When Arens succeeded Ariel Sharon as defense minister, he immediately canceled a Lavi study initiated by his predecessor in December 1981 to examine the possibilities of incorporating Lavi technology with the General Dynamics F-16 airframe. The study had been prepared by General Dynamics, which was in the midst of delivering its first batch of F-16 aircraft to Israel. In fact, it was the government of Israel that requested the study; it was based on Israeli ground rules for its analytical framework.

The study was not GD's first effort to address a partnership with Israel for Lavi production, nor did it represent the first time Israel had broached the subject with the company. Like several other American firms, GD had been visited in 1980 by a senior IAI executive regarding joint production of the Lavi. GD had responded the same year by visiting Israel for a further round of talks.

In providing a preliminary briefing of its study findings on February 15, 1982, the company had insisted that the Lavi involved not only operational considerations, but also those associated with Israel's need both to bolster its economy and achieve self-sufficiency. Moreover, the company stressed that it had maintained a consistent policy of noninterference with respect to the Lavi program and had volunteered information and assistance where feasible. GD's disclaimer was both honest and not surprising; the company could ill afford to alienate one of its most important F-16 clients. The study postulated that Israel Aircraft Industries would remain the overall prime contractor for both the new aircraft and the program as a whole. Nevertheless, by incorporating the F-16 airframe and the American F-100 engine, the Israelis would likely lower their nonrecurring development costs by about 50 percent. Even more interesting was GD's assertion that the Israeli man-hours content of its proposed hybrid, which it estimated to be about 8 percent of the total, might actually exceed the Israeli man-hours total of the Lavi, which relied very heavily on American-manufactured subsystems, despite being advertised as an indigenous Israeli aircraft.

Arens was not impressed by the numbers or the analysis. He was simply not interested in an F-16 or in a hybrid plane. He wanted a product that was primarily, if not solely, Israeli, and did not accept the argument that a program with an American airframe could be more "Israeli" than the Lavi. In fact, he particularly wanted the airframe to be Israeli. Sharon, the former tank commander, may not have been as sensitive to this matter as Arens, the engineer. Whatever the reason, the study was canceled, and the stage was set for a full-court press to obtain American money for an Israeli-dominated development program.

November 1983 was a triumphant month for Arens. With the help of his friends in AIPAC and on the Hill, the Congress passed on November 14 a new foreign assistance bill that reflected virtually in its entirety the original amendment that Charles Wilson and Marvin Klemow of IAI had jointly conceived. Like the text of Wilson's draft amendment, the bill included $500 million in funding for Lavi research and development. Only the amount of money that could be spent in Israel for the project had been reduced, from Wilson's original amount of $300 million to the smaller but still significant total of $250 million. The latter sum did not actually constitute all of the "offshore procurement" (OSP) made available to the Israelis. It was noteworthy, however, because it was "earmarked" for the Lavi. In other words, even if elements within the Israeli military, few of whom were enamored with the project, wished to spend the monies on other programs, they could not do so.

Arens then accompanied Shamir on the prime minister's first visit to Washington since he took office. The atmosphere was exceedingly cordial, a welcome change for Israel after the tensions that had clouded relations between the two countries since the invasion of Lebanon the year before. On November 30, Ronald Reagan signed a memorandum of understanding with Israel that set the basis for a new level of Israeli-American "strategic cooperation." Included among the many commitments that the United States made to Israel was Reagan's support for funding the Lavi development program, thereby ensuring that the Pentagon, whatever its doubts about the plane—and these were legion both in the Office of the Secretary of Defense and in the Joint Staff—would salute smartly and not try to subvert in any way the recently passed legislation.

Ironically, Arens was only briefly able to watch the Lavi funding flow to Israel from his vantage point as defense minister. Begin's resignation had rocked the Likud Party, and Shamir was unable to lead it to victory in the July 1984 elections. By September, Arens had lost his job under the complex terms of the agreement between the Likud and Labor parties that resulted in the creation of a National Unity government. The new defense minister was Labor's Yitzchak Rabin: war hero, former chief of staff, former ambassador to Washington, former prime minister, seemingly former everything. Rabin had returned from political oblivion and, under the terms of the two party agreement, was ensconced as defense minister for the lifetime of the National Unity government. Arens was relegated to the job of "minister without portfolio," which meant that he was too important to be left entirely out in the cold, but could get no closer than the government's front porch; he had no ministry to call his own.

With little by way of administrative duties, Arens was able to act as a sort of government gadfly, particularly on national security matters. He still had considerable clout within the Likud Party, and his relationship with Shamir, once again foreign minister—and prime minister designate in two years' time under the terms of the National Unity deal—was as close as ever. He also sat in the so-called inner cabinet that made key national security decisions. If he had a pet project, it was a fair bet, therefore, that he could push it to fruition. And no pet project meant more to him than the Lavi. It was his creation, and he was determined to make it fly.

2

Something About This Program Stinks

Having been formally given Fred Iklé's charge to head a new Defense Department review of the Lavi program, I received my first briefing on the subject on March 13, 1985. Actually, it was Fred Iklé who received the briefing in his office; he asked me to sit in. I remember almost nothing about it, because I was still consumed by my negotiations with the Canadians. I remember the exact time, as well as day, that I completed the agreement to upgrade our common North American early warning system: 10:00 A.M., on March 17, 1985. It was St. Patrick's Day. Several months earlier, not long after I had been assigned to revive the stalled negotiations, I had been given an additional "tasker": to complete them in time for the president's visit to Canada on St. Patrick's Day. The North Warning agreement was to be one of the highlights of this "Shamrock Summit" with Reagan's fellow Irishman and conservative Brian Mulroney.

It is no easy task to complete a negotiation that has long been on the road to nowhere: positions have already been fixed in concrete on all sides, not only between the countries involved, but within the bureaucracies of both. Particularly when the negotiations involve issues that do not consume the full-time attention of cabinet-level officials, upper-middle-level bureaucrats often forget that they are dealing with their counterparts from another country. Their primary concern is to get their own point of view adopted by everybody else, "coordinated" or "getting everybody to chop" in the jargon. Once all the representatives of all interested agencies have "chopped," negotiators are then often surprised to discover that the foreign delegation with which they are dealing does not want to "chop." They seem taken aback that the foreigners have ideas of their own, often totally at variance not only with the hallowed interagency bureaucratic consensus, but even with many of the positions taken by individual agencies. As a result, the American negotiators have to return to the interagency forum seeking a new consensus, unless the level of interest in the negotiation is sufficiently high so as to prompt decision-making at the highest government levels. Since that was not the case with the Canadian negotiation, my initial task had been to "come up with something different" to break the bureaucratic gridlock. Once that mission was accomplished I was then to get the other side to accept our new position, at least in broad outline, and then negotiate the final agreement. But my new assignment was not only to accomplish the first task by noon on St. Patrick's Day, but to conclude the entire process, to everybody's satisfaction of course.

It wasn't until that pleasant Thursday morning in Quebec, in a rather large room where the Canadian team sat at one end and my people clustered in the other, that we finally reached agreement. Two hours later, overlooking the Plains of Abraham, the agreement was announced, and I looked forward to some quiet weeks of merely managing the hectic day-to-day business of my office.

With nothing to do the rest of the day but enjoy myself, I did just that, admiring the European ambiance of the small French Canadian town. That evening I attended the gala that was the highlight of the Shamrock Summit. The president and prime minister shared the stage as part of the variety-night extravaganza. The only somewhat sour note was struck by the Quebecois nationalist prime minister, René Lévesque, who sat in his box overlooking the proceedings while smoking like a chimney—deaf to the announcement that smoking was prohibited and seemingly blind to the rather large sign nearby that read—in French only, of course—DÉFENSE DE FUMER.

I returned to Washington thinking more about my upcoming trip to Israel—and to India and Pakistan, to which Iklé was also traveling—than about the difficulties that awaited my immersion in the Lavi project. Still, it didn't take me very long to realize that I was in the soup once I began to look more closely at the Lavi program and what lay behind it. Iklé had made it clear to me from the outset that he could not figure out how to stop the program, even though he felt it was a waste of money. He couldn't even prove that it was a waste of money—it was just his instinct, which tended to be accurate. My job would be to prove that once again his instinct had not betrayed him.

Before I began to work on the project I had to discuss it with one more key person: Rich Armitage, the assistant secretary of defense for international security affairs. Although both he and I reported directly to Iklé, Rich stood above me in the bureaucratic pecking order, and well above me in the political one. He had come to the Pentagon as Senator Bob Dole's principal assistant and had established himself, while still holding the relatively low-level office of deputy assistant secretary for East Asian and Pacific affairs, as one of Cap Weinberger's principal lieutenants. One of his first tasks at the Defense Department had been to sack the holdovers from the previous administration. It must have been nerve-racking for many of the Carter appointees to wait for the bald former SEAL, who was still built like a tank and sounded like a machine gun, to show up at their office doors and hand them their pink slips.

Armitage later told me he hated doing the job—the guy is as decent as he looks formidable. A devout Catholic, with eight children, many of whom are adopted, he has a strong sense of right and wrong that surprises many people, and that proved particularly helpful to me as I got further enmeshed in the Lavi business.

When Armitage was promoted to his current position in 1982, everybody—including Fred Iklé—recognized that while Armitage nominally reported to Iklé, he could always go directly to the secretary. In fact, Armitage rarely did so, still further testimony to the kind of person he is—a blunt, honest straight shooter who detests fools and knaves.

Rich was also exceedingly turf-conscious, however, which is why I was apprehensive about raising the Lavi issue with him. He defended his staff's errors against any outside criticism, dealing with them internally at the time of his choosing. Moreover, he would go out of his way to protect their prerogatives against any other staff or office. Not surprisingly, his staff loved him, and he built up a group of devoted followers, some of whom moved with him to the State Department when he became a trouble-shooting ambassador-at-large in the Bush administration.

Iklé's "tasker" to me on the Israeli program meant that I was poaching on Rich's turf—ISA covered relations with all states except those of Europe—and in order to succeed at all I needed his blessing, or at least his benign neglect. To my very pleasant surprise, I found Rich was exceedingly supportive. He had not been a Middle East expert—upon taking the job he told me that he knew there was so much bull on all sides that he would just take each issue on its merits. He proceeded to do so, and soon won the respect and friendship of many leading Israelis and Arabs.

Rich shared Iklé's doubts about the Lavi and was ready to help in any way he could. Ultimately, though, it fell to me to figure out a way to deal with the issue in terms that would clarify once and for all whether the United States should stop carping about the project, or, alternatively, and far more difficult, have the project terminated. I told Rich that my intention was first to obtain the most accurate cost figures available on the project. As a former program analyst for the Congressional Budget Office, I had spent several years evaluating and critiquing military programs whose costs were usually greater than what the military services publicly claimed they were. Often too, the programs did not deliver the levels of effectiveness that the services advertised, so that when compared to alternative program approaches, the service favorites were not worth the money, or, to use the jargon of systems analysts, "cost-effective." Armitage was not a systems analyst. But he recognized the usefulness of my approach. If we flushed out the Israeli numbers, we could get a better sense of the Lavi's inherent value, particularly in relation to the alternative of simply procuring aircraft from the United States.

There was an obvious alternative to the Lavi: the F-16 fighter, produced by General Dynamics. Israel had already acquired seventy-five A and B versions of the F-16 under the 1978 "Peace Marble I" program. It was an F-16A fighter that in 1981 dropped Israel's bombs on the Iraqi Osirak reactor, a daring move that set back Saddam Hussein's nuclear plans for years and, in hindsight, clearly saved the West from facing a nuclear threat as it prosecuted Desert Shield and Desert Storm a decade later.

The F-16 program did not just involve General Dynamics. Its web of subcontractors was spread throughout the United States and beyond. A list of subcontractors produced by the U.S. Air Force in mid-April ran to seven pages, and included nine of Israel's top military systems houses, among them Israel Aircraft Industries. IAI and other companies were also providing additional F-16 subsystems and components as contractors to the Israeli government, which then included the items as government-furnished equipment for the F-16 A/B aircraft that Israel had procured from the United States.

At the very time that I was being drawn into the Lavi controversy, Israel was negotiating a new industrial participation agreement with the United States. On August 19, 1983, Israel had reached agreement with Washington for the purchase of an additional seventy-five F-16s—fifty-one F-16Cs and twenty-four F-16Ds—at a cost of $2.24 billion. These planes differed from American F-16s in a variety of ways. For example, their landing gear tolerated a higher-capacity plane, and they incorporated thirteen separate conventional weapons changes, a variety of software and cockpit changes, and a telebriefing capability. Moreover the F-16s included Israeli-developed avionics, such as an enhanced fire control computer, a central data computer, and the intercom, as well as four classified systems that the Israelis would only describe to American officials as systems A, B, C, and D.

The industrial participation agreement was an adjunct to this "Peace Marble II" sale. As of June 1985, Israeli contractors had received firm orders totaling about $135 million (1981 dollars). The new agreement would give still more F-16 business to IAI and other Israeli contractors.[1]

Procurement of still more F-16s beyond the 150 already acquired, or indeed the purchase of another aircraft, clearly was going to bring yet more work for Israeli industry. Alternatives were not, however, the immediate issue. I had convinced my seniors and colleagues that step number one was to obtain "the numbers," and it was to that task that I turned. It proved to be far more difficult than I expected.

I put in a routine request to the Israeli air attaché at the embassy for cost information on the Lavi. The Israelis were not thrilled. Why did I want the numbers? What was my assignment? The suspicion was endless. And the numbers took some time to reach me. When I finally obtained the Israeli estimates, about ten days after I requested them, they proved to be skeletal in terms of detail. More significant for me, however, was that they were 1982 estimates. I was stunned. It was now March 1985. The program was reputed to be the most important project Israel had ever undertaken. Yet the only estimates that I was sent were three years old. In the United States, estimates, by law, were revised on a quarterly basis, and presented to Congress in documents called Selected Acquisition Reports. Moreover, not only did congressional legislation force the Pentagon to calculate increases in the cost estimates, it also prescribed special remedies for increases that exceeded 15 percent of those estimates.

I phoned the Israeli embassy, only to be told that the estimates we had been sent were the only estimates they had. Our own Pentagon people had nothing more either. It was incredible, and I didn't believe for a minute that the Israelis had nothing more recent. They were hiding something. And I reasoned that if they had something to hide, it couldn't be too good.

Israel has never been shy about trumpeting its successes, and sometimes about trumpeting activities that it wishes people to think are successes. When Armitage asked me one afternoon as I passed his doorway how things were going, I told him what was happening with the numbers. I told him that I'd approached the project with an open mind, and in a way wanted it to succeed. But even though I'd been on it only about two weeks, the lack of information bothered me. I thought the Israelis were stonewalling. I remember the words I used, because I repeated them several times afterward: "Something about this program stinks."

Armitage was not surprised when I told him. Neither was Fred Iklé. The three of us reviewed the matter on March 26 with Lieutenant General Phil Gast, the head of the Defense Security Assistance Agency. It was DSAA's task to supervise all foreign military sales, and therefore to administer the transfer of matériel to the Israelis in support of their Lavi program. I had first met and worked with Phil when he was directing operations for the Joint Chiefs of Staff during the Falklands War three years earlier. At the time I was responsible for overseeing our resupply of British forces. I had found Phil to be honest to a fault, not mincing words when he was unhappy about something. Phil was not at all thrilled by the Lavi program. He felt that the Israelis were taking America for a ride, and was frustrated that his agency had the task of providing the Israelis with the wherewithal to carry their development program forward.

I had still not worked out in my mind just how I was going to get the Israelis to reconsider, or at least review, the Lavi program. Iklé, Armitage, and Gast were all expecting me to come up with something, but the meeting did not lead to any obvious ideas. The challenge for us all was to get to the bottom of the matter and to find out how badly things had gone. Could the program be saved? Should it be saved? We didn't think so, but no one really knew for sure. Moreover, even if the program was best killed, how could it be? Israel was a juggernaut in Washington in those days. Reagan's November 1983 agreement with Shamir had lent an entirely new tone to relations between the countries. Prime Minister Shimon Peres, a Washington favorite, was a devotee of high technology and a declared supporter of the Lavi. Rabin, who had criticized the project while out of office, had now come out in its favor. He too was very popular in Washington, even more so than Peres. And there was Arens lurking in the wings, ensuring that no one deviated from the line he had established as ambassador and defense minister.

To say I was apprehensive was to put matters very mildly. I was really scared. To begin with, it was clear to me that the new strategic relationship that Reagan and Shamir initated would make it far more difficult to pursue policies that put the countries at cross-purposes, as any effort to challenge the Lavi clearly would. Though presidents and their staffs always recognize that all such agreements limit the future flexibility of the United States, they often find it difficult to envisage concrete examples of such limitations. Yet it was precisely within the constraints created by this agreement that I had to function.

Moreover, I had heard of the virtual brutality with which the Israelis dealt with those who opposed them, especially if they were Jewish. The chief rabbi of England, Sir Immanuel (later Lord) Jakobovits, a longtime supporter of Israel, had been bitterly defamed in a whispering campaign launched by the Israelis after he criticized their 1982 Lebanon operation. No one could prove that any Israeli had actually said anything for the record, and Jacobovits, whom I knew from my student days at Oxford and my membership in the chief rabbi's University Chaplaincy Board, never responded to the slurs and whispers. He was too much the gentleman to do so. Nevertheless, my friends in England all confirmed that they had heard the attacks, which were also spread in South Africa, Australia, and the rest of the English-speaking Jewish world. Since the Israelis played hardball, and played it personally, I figured that I would soon be on their hit list.

Whatever my apprehensions about the Israelis, I had a more immediate problem: my boss. Iklé was pressing me for some material to take with him on his trip to Israel, which was scheduled for late April. Early in the first week of April he asked me to find the analytical basis for dissuading the Israelis from moving ahead with the program. I did not get the sense that Iklé really expected me to succeed this time.

It was by now very clear to me that the Israelis intended to block my efforts to obtain cost data from them in time for our trip. They hoped that we would arrive in Israel with nothing other than vague concerns about the project. Iklé would make a little bit of noise and we would then all go away. The project could proceed as planned, with no further perturbations from nosy Americans. And I would disappear into the Pentagon woodwork from which I had so inconveniently emerged.

Since the Israelis were obviously not going to be forthcoming about their costs, we had to generate estimates of our own. I recognized, of course, that whatever estimates we put forward were likely to be attacked by the Lavi's proponents, both in Israel and in the United States. Therefore, I intended to use the figures we produced to elicit the real numbers from the Lavi program management. If we could infuriate them sufficiently, we could flush them out, find out what costs or cost factors they were hiding, and, on the basis of evaluating their own true calculations, expose them to public debate in Israel—and, for that matter, in America. If, as I anticipated, those costs were much higher than the figures the Israelis had thus far made public, we might then make so strong a case against moving ahead with the Lavi that we ultimately could get the Israeli government, if not our own Congress, to reconsider the wisdom of pressing ahead with the airplane.

I outlined my approach in a meeting with our ambassador to Israel, Sam Lewis, that Iklé hosted in his office on April 1. Sam, a career foreign service officer, was nearing the conclusion of a historic and highly successful seven-year tour in Tel Aviv. He knew the Israeli mind-set about as well as anyone, and he was skeptical that we could get anywhere with the Israelis, though he did not discourage our efforts.

What counted for me, however, was Iklé's attitude, and Fred agreed to my approach. That only meant that I had to get moving very quickly, however; the trip to Israel was less than three weeks away. Passover also was virtually upon me. The holiday began on the eve of Thursday, April 4, when the family celebrated the first of two seders, the second being held the following night. Passover meant that, at a minimum, I would be out of the office for four days, the holier days of the festival (one of those days was Saturday, when

I never came to the office anyway). Orthodox Jews, following scriptural injunction and rabbinic prescription, treat the first and last two days of the holiday as they would the Sabbath: we do no "work," with work defined in the very special sense that forbids everything from watching television, to telephone conversations (or any use of electricity), to traveling, to writing, to more commonly understood meanings of the term such as cleaning or mowing the lawn. The prohibition during these four days is relaxed in one respect only, namely, cooking for the holiday is permitted, though that reprieve did little for my work on the Lavi. To be sure, one can violate the holiday, or indeed the Sabbath, when life is endangered. Nevertheless, although I did work in the one government department where lives were frequently at risk, and although from an Israeli perspective the fate of the Lavi was a life-and-death matter, it did not seem to me that cost analysis merited violating the holiday. I was simply going to be out of commission for four days in the middle of April, and there was little I could do about that.

Since Passover is actually an eight-day holiday, some readers might wonder about the other four days. Aren't they holy too? In fact, the rabbis of the Talmud and subsequent jurists frowned upon work during those other days as well (those days are termed *chol ha'moed,* the unsanctified holiday). Nevertheless, work is not absolutely prohibited, and urgent matters can be dealt with in almost the same manner as during any other day of the week. The Lavi might not be a life-and-death issue for me, but it was certainly an "urgent matter."

My personal religious concern was compounded by the practical problem that I was in charge of a rather small office. In fact, the office was really of my own creation. Soon after I had gone to work for Fred Iklé, I had convinced him that although I was his special assistant, I needed support in order to provide him with the analytical research, particularly on weapons programs, that he desired. Moreover, Iklé considered me to be his prime lieutenant in managing the planning phase of the Pentagon's annual budget development process, for which he, as under secretary, was responsible. He soon put me in charge of producing the secretary's *Defense Guidance* (now called the *Defense Planning Guidance*), which was, and still is, the department's primary program-planning document. To carry out this additional task I needed still more staff support, particularly as I was also being assigned ever more complex trouble-shooting tasks as well. Nevertheless, my enlarged staff remained a very small one, in Pentagon terms, consisting of no more than twenty people at any one time. I managed to carry out my duties only by super-imposing this staff upon the work of special task forces and steering groups drawn from the military services and various divisions within the Office of the Secretary of Defense. Thus, the number of people involved in producing the *Defense Guidance* in any given year was probably ten times as large as the team in my immediate office.

In order to carry out my plan to analyze the costs of the Lavi, I recognized that I would have to adopt a management formula simi-lar to that for producing the *Guidance.* Iklé was neither willing nor able to double my staff, yet I estimated that I would ultimately require at least twenty more people to carry out the analysis properly. With the assistance of my senior deputy, Dick Smull, an air force colonel who had come to me from the planning office of the deputy chief of staff for research, development, and acquisition (RDA), I was able to pull together a team of brilliant cost and systems analysts from the Air Staff's planning office and from the secretary of defense's Office of Program Analysis and Evaluation (PA&E).

I was fortunate to have strong support and sympathy from the people in charge of both units. The general who headed the Air Staff's office of operational requirements was Major General Mike Loh, a blunt, forthright aviator who would later rise to become the four-star vice chief of staff and then head of the Air Combat Command. Mike and I got along extremely well; we both knew exactly what needed to be done, and he made sure that the air force was ever ready to do its share to support my efforts. The director of PA&E, who served in that post throughout the Reagan and Bush administrations, was my former boss at the Congressional Budget Office, David S. C. Chu. Dave Chu and I had always had correct, though not warm, relations. A brilliant systems analyst who had graduated Yale *summa cum laude,* was a decorated army officer, and returned to Yale to earn his doctorate in economics, Chu rarely took political risks and generally kept his own counsel. But Dave recognized that the case of the Lavi was in some ways an analysts' dream. It offered the analyst the very rare opportunity to employ rigorous cost methodologies and estima-tion to affect not only national policy, but international policy and affairs. Like Mike Loh, Dave was willing to assign to my team some of his top specialists, even though it meant pulling them away from other pressing needs within his office.

On April 2, 1985, I received some assistance from Phil Gast's office, though at the time I was not exactly sure how I would make the most of it. It came in the form of a memorandum from Glenn Rudd, Gast's deputy, who was then acting director during Phil's temporary absence, to Iklé, regarding changes in the congressional Lavi legislation for fiscal year 1986. Glenn was a soft-spoken professional who, like Gast, loved his job but hated to be exploited. While he was by no means anti-Israel, he took a very hardheaded view of its unceasing and creative requests for U.S. government assistance. The Israelis tend not to like people who deal with them in a matter-of-fact way; that is a manner they prefer to arrogate to themselves. So they at times said things that were less than nice about Glenn Rudd. But it comes with the territory when dealing with Israel, and Glenn never let the innuendo bother him very much.

Glenn's memo noted that the Congress was broadening the scope of its Lavi-related assistance to Israel. Both houses of Congress were inserting language that awarded Israel funds to spend on the Lavi (in the language of the House Foreign Affairs Committee) "if the Government of Israel requests that funds be used for such purposes." As DSAA saw it, Israel's supporters were attempting to "insure that Israel clearly has the option of either spending or not spending the earmarked funds for the Lavi on the Lavi program." No doubt Israel's supporters had no intention of curtailing the Lavi program. But the change meant that I could now seek the termination of the Lavi without any financial loss to Israel. This development was to prove critical eighteen months later as I sought to make clear to many Israelis that there was an "opportunity cost" to the Lavi; the money could be spent on more deserving and financially starved Israeli military programs.

The Rudd memo was actually prepared in anticipation of a meeting that Fred Iklé and I had scheduled with Senator Rudy Boschwitz the following day. Boschwitz, a good-natured and exrtremely successful Minnesota businessman, had been elected to the Senate on Reagan's coattails in 1981. He had been very active in Jewish affairs prior to his election, and he remained a staunch friend of Israel while he was in office. Fred Iklé reasoned that he needed Boschwitz's neutrality, if not support, as he pressed ahead with his effort to ground the Lavi. Iklé expected some problems from Israel's Democratic friends; it made good political sense to bash the administration on an Israeli-related issue. He could not afford to alienate Republicans as well. Members of Congress hate to be "blindsided." Information is their coin of the realm, and they, and their staffs, relish their access to administration decision-making before it is actually finalized. Accordingly, we went over to Boschwitz's office to give him a "heads-up" that we were going to surface the Lavi issue when we visited Israel.

Boschwitz was preoccupied with other matters. When we arrived at his office in the Hart Building, Boschwitz had been discussing price supports with some disgruntled farmers from his home state. He explained all this to us when we were ushered in to see him. Iklé tried very hard to be interested. We promised to keep Boschwitz informed about developments on the Lavi, and left his office less than a half hour after we arrived. The episode presaged the nature of congressional interest throughout the affair: a desire to know what was going on, but no fundamental involvement in the issue.

During the next few days I didn't have much time to think about Congress at all. I now had the people I needed to produce our cost estimates, but I still needed a methodological strategy. I simply could not generate costs out of the air; the Israelis would then be able to dismiss our efforts with a wave of the hand, which is precisely what they hoped to be able to do.

After numerous discussions with my team, and most particularly with a young air force captain named Linda Hardy—one of the most brilliant and perceptive cost analysts I have ever met—we decided upon two very different approaches to the problem of estimating the Lavi's costs. One was generated by the Office of the Secretary of Defense (OSD), primarily by Dave Chu's analysts at PA&E. It derived from a comparison of originally advertised cost estimates for the development and production of the three of our top-of-the-line fighters, the Air Force F-15 and F-16 and the Navy F-18, with the actual or final cost of these programs. PA&E determined what it termed development and production growth ratios for the fighters and applied similar ratios to the Lavi's official cost estimates, thereby arriving at a new projected cost for the Israeli plane.

The other set of estimates was produced by the air force, with Linda Hardy doing yeoman's work under the direction of Colonel Mike Foley, who worked for General Loh and was a distant relative of then Speaker of the House Tom Foley. Linda and her fellow analysts first assembled relevant factors from the air force's vast storehouse of analytical material on aircraft cost estimation, particularly some very sophisticated databases for the three fighters in the PA&E analysis, the F-15, F-16, and F-18, as well as for the navy's F-14 fighter, the air force's A-10 attack plane, and the navy/air force A-7 attack aircraft. The team examined what it termed "interval times" between initiation of each aircraft project and production start-up—in other words, the time it took to develop the plane, and then the time needed to produce the entire run of aircraft. The team then related the costs of development and production to the two time intervals for each of the American aircraft. Finally, applying the mathematical formula that expressed the time interval/cost relationship for American planes to the projected time intervals for the Lavi program, the analysts would generate an estimate for both the Lavi's development and production cost.

While the methodology probably seemed arcane to all but devotees of cost analysis, the results were nothing short of startling. The OSD analysis projected a development cost of $3 billion, twice the original Israeli estimate and $500 million more than the most recently revised figures that the Israelis had been prepared to release to us. The difference in production cost estimates was even greater. OSD estimated those costs at $13.9 billion, more than double the upwardly revised Israeli estimate of $6.8 billion.

The air force analysis that Linda Hardy and her colleagues put together made the OSD work look modest by comparison. The air force team projected development costs of no less than $10 billion, or more than $3 billion in excess of what the Israelis had estimated the plane would cost them (and us) to produce. Surprisingly, however, the production cost estimates were more modest than those of OSD. Still, at $12.9 billion, they exceeded the Israeli revised estimates by over $6 billion.

As we put our briefing together, I realized that my material was nothing short of explosive. The issue became one of packaging: how could I maximize the impact of what I was about to reveal to what I knew would be an incredulous Israeli defense establishment? I decided that we would offer the Israelis a briefing, and the briefing slides would all be unclassified. Strictly speaking, there was no reason why we should have classified our material, since there was nothing secret about it. But the Pentagon and other agencies have often classified their documents for reasons other than national security in its narrowest sense. Sometimes those reasons include a desire to keep the public in the dark about some material that might be embarrassing to the administration in general, or one of its agencies in particular. In our case, the only embarrassment would be to the Israelis, and I had no desire to protect them from any of it.

I also decided that we should present both sets of estimates, which had been produced independently of each other, with the argument that even the lower estimate, that of OSD, devastated the Israeli case. Finally, I felt that we needed to couch the entire briefing in nonthreatening terms. I wanted to isolate the Lavi's proponents, not alienate the government and people of Israel. Our case would be based on efficiency; we desired to make the most *for Israel* of the money that U.S. taxpayers had provided to that country.

Although the staff put together the details for the briefing, I decided to prepare it on my own. I was determined that our message should be as friendly and as helpful as possible, and at the same time sufficiently substantive and analytical so as to avoid being dismissed out of hand by the Lavi's proponents. Most important, I decided that I had to raise the issue of alternatives to the Lavi; in that way not only would the internal Israeli debate focus on whether or not to continue with the program, but it would also address the fact that, in the F-16, F-18, and other American aircraft, Israel had a viable way of acquiring top-flight aircraft "off the shelf," without incurring the expenses of developing an entirely new system.

As I neared completion of the briefing, I obtained Iklé's permission to inform Israel's primary organizational supporter in Washington, the American Israel Public Affairs Committee, of our intentions on our forthcoming trip. Both he and I recognized that AIPAC was probably fully aware of what we were planning; we also knew, and indeed hoped, that if the Lavi became a major issue in Israeli policymaking circles, it would assume almost equal prominence in the American press, on Capitol Hill, and on the agendas of the major American Jewish organizations. AIPAC was certainly the best-known and most capable of those Jewish organizations that operated in Washington. Moreover, it was probably at the height of its influence in the mid-eighties, given President Reagan's visceral support for Israel and the exceedingly warm relations between George Shultz's State Department and the government of Prime Minister Shimon Peres. It was obvious to both Iklé and to me that AIPAC was the first place to which we should turn in reaching out to American Jewry.

I was under no illusions about co-opting AIPAC to our cause.

I did hope, however, that by opening a line of communication early in our effort, we might avoid the rancor that had colored AIPAC's disagreements with other administration policies, such as the bitterly disputed 1981 sale of AWACS aircraft to Saudi Arabia. In fact, we needed a quiescent AIPAC for another reason: we were not sure that the State Department was ready to go along with us. Shultz's friendship with the Israelis had made him their most powerful advocate in the U.S. government. His rivalry with Caspar Weinberger, whom the Israelis distrusted and who was, quite unfairly, despised by the organized Jewish community in the United States, further complicated any effort by the Pentagon to push through a policy that might be interpreted as anti-Israel. We needed AIPAC's quiescence so as not to arouse Shultz's suspicions.

I did have some reason to hope for a friendly dialogue with AIPAC, since I had remained on good terms with its talented executive director, Tom Dine. I first got to know Tom in the mid-seventies when I was a very junior defense analyst at the Congressional Budget Office and he had become director of the Senate Budget Committee's national security task force. A dyed-in-the-wool liberal, and proud of it, Tom had previously also served in the Peace Corps in India. From the Hill he had moved to the Brookings Institution, where he worked on defense issues and hoped for an appointment with the Carter administration. He received several offers, but none to his liking, and then was offered the AIPAC job.

Tom had asked me to join him as his deputy at AIPAC. I had politely declined; my interests lay elsewhere. But I had no doubt that he would succeed in his new job. He was as nice a person as he was talented; he was simply pleasant to deal with—a trait that goes a long way in Washington. I had not realized—until he told me about his impending move from Brookings—just how committed to Israel Tom Dine really was. He stemmed from a family that was prominent in Cincinnati Jewish circles (and indeed, beyond—his brother Jim is a well-known painter) and, like many who hail from America's capital of Reform Judaism, is a Reform Jew. He visited our home on more than one occasion, and knew, of course, that I was Orthodox. He never seemed uncomfortable about my Orthodoxy any more than I was uncomfortable about his background.

I mention all this because Tom was forced out of AIPAC in the summer of 1993, supposedly for having made some disparaging remarks about Orthodox Jews to an Israeli reporter. Tom would not deny responsibility for the remarks once they were published, and, I am told, he apologized to the entire AIPAC staff. In fact, it is more in Tom's character to take the heat for a mistake he may have made than to have made that mistake in the first place; he was always a man who combined personal integrity and sensitivity to others' points of view.

Tom had been a receptive audience for Moshe Arens's message, and not only because of his own strongly pro-Israel views and the nature of the organization he led. Tom had long felt that Israel needed to be as self-reliant as possible. The Arens proposal pointed in that direction. It would allow Israel to continue to strengthen its still fledgling aircraft industry, and, with a massive influx of dollars, telescope what otherwise might be a decades-long process. Israel would not only be able to build an infrastructure that would ensure it the wherewithal to maintain its qualitative military edge over its adversaries, it also would be an even more attractive vendor on the world arms market—possibly even to the United States itself. Thus, for Tom, the Lavi proposal was a natural.

While I recognized that Tom had a constituency to represent, I remained convinced that Tom was just too good a defense analyst to swallow the Lavi story whole once I put my case to him. Moreover, he had always been one of my strongest supporters when I had made my own analytical presentations to the budget committees while I was still at the CBO. I therefore looked forward to our meeting with a degree of what Washington hands call "cautious optimism."

The encounter did not get off to a great start. Tom was forty-five minutes late. I am not exactly prompt myself, but after a half hour I was steaming. Still, once he arrived, we managed to have a friendly enough conversation. I told Tom that I was working on the Lavi and that I was having an increasingly tough time keeping an open mind about the project. Something was wrong, and I hoped the Israelis would be forthcoming enough so that matters could be straightened out as quickly as possible.

Tom was noncommittal. But he did not seriously try to dissuade me from carrying out my job—that would come later, from other people, including some in AIPAC, but never from Tom. He did warn me that my involvement could cause me personal problems, and I soon learned that he was not overstating that danger. I told Tom that I'd keep him informed after I returned from my trip, and I had every intention of doing so, but I left him with the feeling that I would be very lucky if AIPAC displayed no more than token opposition to our effort.

As we approached the date of our departure, Iklé and I continued our round of "heads-up" meetings on the Lavi. We did not get too much sympathy from Bob Kasten, the Republican chairman of the Senate Appropriations Foreign Operations Subcommittee, and even less from Jim Bond, who had worked so closely with Moshe Arens to obtain initial congressional funding for the Lavi. We didn't do much better with Kasten's Democratic counterpart, Dan Inouye, another strong supporter of Israel, who was cordial enough, but was very cautious about saying anything that we might construe as being critical of the project. There was not going to be much support coming from the Hill; that much was clear. I could not help but recall Sam Lewis's reaction to my strategy when I had briefed him some three weeks earlier; he gave it at best a 5 percent chance of succeeding. And it was with that somber prediction in mind that I began my first visit to Israel in nearly a decade.

3

Getting Down to Business

I had visited Israel several times prior to the first of my Lavi-related trips. My first visit was in conjunction with my younger brother's bar mitzvah, in August 1967. My family had not been planning a bar mitzvah in Israel; mine had been held in Manhattan. But my brother Josh had many years earlier—he was perhaps four at the time—told my great-uncle, the leading Hasidic rebbe in Philadelphia, that he wanted to have his bar mitzvah at the Temple in Jerusalem. At the time this seemed out of the question, since East Jerusalem was firmly in Jordanian hands. But when Israel triumphed in the Six-Day War, my great-uncle, who by then had himself moved to Israel, recalled Josh's "prophecy" and offered to host the entire family in Israel. It was a marvelous gesture, but for a variety of reasons, I did not find Israel to my liking.

I returned to Israel a few times in the early seventies, when I was able to benefit from cheap air fares from England, where I was a student, and warmed slightly to the place. My last visit to Israel took place in early July 1976. My wife and I, and our oldest—and at the time only—son, Chaim, had spent a week in Jerusalem. I remember being waked by the operator on the morning of July 4 and told to turn on my television set. It seemed a most unusual request—actually, it was an order—but I meekly complied, and we discovered that Israel had just liberated the Entebbe hostages. That day we were treated to some aerial acrobatics over Jerusalem by the Israeli air force, and we joined the Israelis in basking in the success of that remarkable and daring operation.

Although I was part of a rather small team that Iklé was taking with him on a military airplane, I flew to Israel alone. Iklé's aircraft was taking off on Saturday, but I do not travel on the Sabbath. In order to be available for work on Sunday—the first day of the Israeli work week—I had to leave Washington the previous Thursday evening. I spent the Shabbat in the town of Rehovot, which is best known for housing the famed Weizmann Institute of Science. Rehovot is also a reflection of the cultural schisms that lie just underneath the surface of the Israeli melting pot: it is a haven for English-speaking Jews—American, British, Canadian, South African, Australian. I was staying with my friends Neal and Miki Hauser, who had made *aliyah*—that is, had emigrated—from Washington several years earlier. *Aliyah* literally means "go up," since, from a Judaic perspective, one always goes up to Israel, which is spiritually the highest place on earth. (Israel does not have a monopoly on this particular notion. At Oxford we always spoke of people's "coming up" to Oxford. If a student was rusticated, or punished, the student was "sent down" to wherever he or she came from.) When I accompanied Neal to shul—synagogue—on Shabbat, it was as if my entire life were packed into a single district of this small Israeli town. I met people on the way to, from, and in the shul whom I had known as a high school kid in Brooklyn, as a student at Columbia, as a junior-year student in London, as a camper in Pennsylvania, as a graduate student at Oxford. In fact, during my two block walk to shul on Friday night, I stopped five times to meet and greet long-lost friends and acquaintances of years past. Invariably they wondered why I was in Israel; they couldn't believe that an Orthodox Jew was really on U.S. government business.

Shortly after I checked into the Tel Aviv Hilton late Saturday night, I began to wonder what I was doing in Israel. Earlier that evening, when Shabbat had ended, Neal and Miki had kindly offered to take me into Jerusalem. I figured that I would probably not have a chance later to see the city at night—though we were scheduled to meet in government offices during the day—so I gladly accepted. On the way up the Jerusalem road I was struck by the sight of hundreds of flickering lights in the western valley below the city. I had "discovered" Ramot, the sprawling Jerusalem suburb that had sprung from the ground like a giant mushroom since my last visit to Israel. Ten years earlier, where Ramot now stood there had been a barren

valley. Now it was a thriving, and still growing, bedroom community for people who worked in Jerusalem. It was my first exposure to
the explosive growth that Jerusalem had undergone in the 1980s.

By the time Neal and Miki dropped me off at the Hilton, it was
nearly midnight, and I was dog tired. Still, I needed to have some
laundry done, and I phoned down to the laundry to ask it to take
some shirts. The lady at the other end of the line was obliging
enough, until she asked for my room number. Three twenty-four,
I said.

"No, you are in four thirty-two."

I was stunned. Maybe I had not got over my jet lag. Maybe I was
still reeling from the remembrances of things past that Sabbath in
Rehovot had evoked. Maybe I was overwhelmed by being in Israel
again. Maybe I was just plain tired; by now it was past midnight.
I looked at the telephone. The number on the receiver was 324.

"Miss, I'm looking at my telephone, and the number says three-
two-four."

"I emm looking et my computayr, you are in forr—thayr—
tee—too."

"Wait a minute, I'll look at my door." I ran to the door. I thought
I was going nuts. I breathed a sigh of relief when I saw the three
numbers on the door. "I just looked at the door, and it says three
twenty-four."

"You are in forr—thayr—tee—too."

This was getting frustrating. It brought back memories of some
of the Israeli kids whom I had known in high school, who were
impossible to deal with. Maybe Israel hadn't changed that much.

"I'll tell you in Hebrew. I'm in *shalosh esrim ushnaim.*"

"I don't kerr what language you use. You are in forr-thayr—
tee—too. You want me to send laundry or no?"

By now I was about to burst. "Lady, my name is Zakheim and
I'm in three twenty-four."

"Sayr, your name is Cohen and you are in four—thayr—
tee—too."

I hung up and went to bed, trying not to think of what else
would befall me on this trip. Befall it did.

Under Secretary of Defense Fred Iklé and the rest of the team arrived
as scheduled on Saturday. There had been some, but not much,
Israeli press comment on our visit. One of the more interesting commentaries appeared in the center-left Hebrew-language paper *Davar*

the day before Iklé arrived. Entitled "Make War, Not Lavi," the lengthy analysis began by linking Fred to the chairman of Northrop, an outspoken Lavi opponent. The article asserted that Iklé was particularly concerned about the Lavi's impact on the American defense industrial base, of which Northrop was a critical component. More important, however, the article went on to say that criticism of the Lavi should not be off-limits for the Pentagon, since there were so many Lavi opponents in Israel, among them Major General Dan Shomron, who later was to become chief of staff. Significantly, we were not scheduled to meet with Shomron at any time during our visit. The article then suggested that Iklé would find an "attractive listener from the past—Yitzchak Rabin . . . [who] has been waiting for six months or more for the American, who will not take the hint and relieve him of the burden of the plane."[1] This was news to me, though we were all aware of Rabin's vocal opposition to the project before he had become a minister.

Israeli radio also carried what proved to be two rather portentous interviews just a few days prior to our arrival. In one, an adviser to the finance ministry on defense matters indicated the doubts that had begun to creep into the ministry's most senior circles. "We know today," he stated, "that the development cost will go approximately to $2.5 billion, when the original plan, when it was presented to the government in the first place, was about $700 million for development; we know today that the price of the aircraft will be more than the F-16." The expert went on to say that he thought Israel needed the planes, but it had to put a price cap on them.

The second interview offered, in the starkest of terms, the other side of the Israeli coin. Former defense minister Moshe Arens claimed that our upcoming visit to Israel was a result of "some curiosity in the United States" and expected that Fred Iklé would "come away impressed." He was certain, he added, that while "some of the criticism . . . leveled at the Lavi in Israel cannot produce positive effects in the United States . . . the program sells itself."

Arens then went on to voice two controversial premises underlying his—and IAI's—assessment of the program's prospects. First, he believed that despite American vetoes of Kfir sales to South America, the United States would not block Israeli sales of Lavi to third countries, even though the Lavi incorporated significant American technologies and would compete with American manufacturers for those third-market sales. "If opportunities do arise in the future for exporting the airplane," he asserted, "I do not foresee any difficulty in U.S. policy not permitting us to make these exports, especially because of the great participation of American industry in the airplane."

Arens's assumption that we would permit the sale of the Lavi to other countries (he also harbored the belief that the Lavi might replace the American A-10 ground attack plane) was far off the mark. Even if the Lavi did come into being, we had absolutely no intention of letting it compete with our F-15s and F-16s. America was very generous to Israel, but our generosity had limits. Despite, or perhaps because of, his years in the United States, Arens simply did not see that he was living in a dream world when it came to his beloved airplane's prospects outside Israel.

Arens's second premise was equally revealing—and wrong. In response to the statement that the Israeli officers were worrying about the impact of the Lavi on the overall Israel Defense Forces budget, Arens said that "some of the criticism is not really founded on full knowledge of the facts. American military assistance to Israel has been generous in the past few years, the funds of the Lavi have been specifically allocated to the Lavi by the administration and by the United States Congress, and I don't think that they in any way decrease the flexibility for purchasing of systems that we have, *except insofar that of course the purchase of any system means that you have less money to spend on other systems*" (my emphasis).[2]

Arens was obviously not prepared to face up to the reality of the Lavi's impact on the defense program. Of course it removed Israel's flexibility. When Arens could state, as he did, that the Israeli air force was an important component of the IDF, that fighters were an important component of the air force, and that the Lavi was Israel's most important fighter, he was in fact arguing that, whatever the cost to budget flexibility, the Lavi had to be procured. And it was precisely that argument that frightened Israel's military leaders more than any other, as I was soon to learn firsthand.

Our group held its first full meeting on Sunday morning in Fred Iklé's hotel suite at the Hilton. The team included Phil Merrill, publisher of *Washingtonian* magazine and Fred Iklé's dollar-a-year counselor at the Pentagon; Scott Redd, his military assistant, a brilliant navy captain who achieved flag rank as an admiral a few years later; and other experts on the Middle East and South Asia. Iklé led the discussion, mapping out for us our approach to the Israelis. We were to be as polite and as disciplined as possible. We should avoid every temptation to let our briefing degenerate into a shouting match. One of us pointed at the ceiling; it was a silent signal that the room might be bugged. We had no proof, of course, and Israel is a friendly country. But why take chances? The conversation thereafter took on a rather elliptical tone.

If any Israelis were listening, they certainly did not let on as I made my presentation. I had prepared a longer and shorter version of my material; the shorter version was for the ministers, the longer version for the staff. Our first stop was the defense ministry, known in Hebrew as Hakirya, "the village," because it is actually a collection of buildings large and small, clustered around a central building with a tall communications tower.

We began by briefing the ministry's Lavi team, headed by its program manager, Menachem Eini. Eini was an Iraqi-born fighter navigator who had been shot down and captured by the Egyptians during the first stages of the 1970–72 War of Attrition and was returned to Israel after the Yom Kippur War. With the conviction that everything he did was absolutely right, with a temper that he could barely keep under control, with his quick smile and warm personality, Eini was very much still the fighter jock.

Eini loved the Lavi. If the plane was Moshe Arens's baby, it was Eini's girlfriend. His emotional attachment to the project masked what we soon perceived to be an absence of tough managerial skills. His job was to oversee the project, to ensure that the ministry, not the contractor who built the plane, was in charge. In this regard Eini was not a smashing success.

It was primarily to Eini that I addressed my briefing. I opened by articulating our three primary goals: to preserve Israel's industrial base; to maximize the employment of skilled Israeli labor in its aircraft program; and to promote efficient use of offshore foreign military sales funds. Only the third goal hinted at what was to come; there was no way that Eini could object to the first two.

The next slide went to the heart of the matter, however. We outlined our apprehensions about the Lavi program. First, even based on Israel's own (and in our view outdated) estimates, its costs had risen so much that we were concerned about its affordability. Second, we outlined a series of technical issues. We were concerned about Israel's ability to integrate the plane's avionics system, which would incorporate subsystems from several manufacturers in different countries. We worried about the performance and cost of the flight control software. Lear Siegler, an American company, was producing this software for both the Israeli Lavi and the Swedish Gripen fighter. We had heard of problems in both cases, and we wondered whether Lear Siegler's engineering resources were sufficient to allow it to perform to standards in both cases. We were also concerned about the plane's weight. We expected potential weight increases as the system "matured," that is, as the engineers produced their various subsystems. Weight increases are endemic to new aircraft developments; they also drive up costs.

We were concerned about scheduling. Would a prototype be ready as promised in September 1986? Would the integration of Lavi avionics and the completion of software take place as planned?

Finally, we raised the question of shifting more production work to the United States, a subject that was sure to strike a raw Israeli nerve. The Israelis wished to build as much of the Lavi as they could. To the extent that they shifted work to the United States—we suggested the engine, the wing, and the tail—they would be losing jobs and, even more important, their ability to produce a plane independently.

I listed all of the foregoing concerns on but one slide. Eini and his team listened impassively. It was when I turned to the cost estimates that the Israelis hit the roof. I thought Eini would have a heart attack when I flashed the U.S. Air Force estimates on the briefing screen. The lower OSD cost estimates were hardly more reassuring. Eini's dark countenance took on a distinctly red hue. I stressed that these estimates were based on our own extrapolation of the limited data that the Israelis had provided to us. We needed Israeli numbers, which, I was sure, would result in lower estimates.

Having demonstrated our figures, I turned to the next touchy issue, that of alternatives to the Lavi program as currently constituted. The first of my final three slides dealt with the fact that a very large proportion of both the Israeli and American contractors performing work on the Lavi also worked on the American F-16 fighter. The inference was obvious: Israel would not necessarily lose much business if it added to its American-built F-16 force. I then laid out the alternatives that confronted the Israelis.

- They could continue the Lavi program.
- They could buy an American-made aircraft.
- They could coproduce an American plane with either an Israeli or American firm acting as the prime contractor.

The concluding slide reemphasized that the Lavi was costly and that there was a need to examine alternatives. I suggested again,

however, that we needed to protect the Israeli economy and indus- 🗡
trial base, and that therefore it was best that we and the Israelis
work together to find the best overall solution.

We were scheduled to brief Defense Minister Rabin next, and
Eini had little time—or inclination, it seemed—to comment directly
on our findings, other than to indicate that he rejected them in toto.
We proceeded to Rabin's relatively small office in the main building
of the defense ministry. I was thrilled to be meeting the hero of the
Six-Day War, who had been the subject of popular Israeli victory
songs I had learned as a boy.

The man who greeted us was of medium height, balding, tanned,
but clearly fatigued. His jowls reminded me, as if I needed reminding, that here was a man old enough to be my father. I was struck by
Rabin's gruff low-pitched voice, his tendency to frown rather than
smile, and his chain-smoking, which by that time was a rarity in
Washington. He spoke slowly, deliberately. His English was quite
good, far better than when he had first come to Washington as
Israel's ambassador to the United States.

Rabin was polite enough, and was a very good listener. I showed
him some of the slides at the edge of his table. We were interrupted
by his tea lady, who wore combat boots and looked as if she could
vanquish the entire Arab world just by staring at it. Rabin asked her
to come back later. She gave him her Medusa look, and the defense
minister of Israel, the war hero, backed off. The tea was served.

Rabin was joined by Menachem "Mendy" Meron, the ministry's
director general and Rabin's right-hand man. Mendy, a retired army
major general, had capped his career with three years as a highly
popular defense attaché in Washington, where he became quite close
to Arens. A rotund fellow with a good-natured air, Mendy understood very well that we posed a serious threat to the Lavi, and came
across as a forceful advocate of the project's continuation.

I gave Rabin and Meron an abbreviated version of the briefing.
Rabin did not ask too many questions. We told him that we were
going to brief Israel Aircraft Industries, as well as Prime Minister
Peres, among others. We arranged to meet him again the next day, at
the conclusion of our visit. I had no idea whether he would support
our effort to reevaluate the Lavi, despite the press reports about his
continuing doubts about the program. I decided as I left that I really
liked the man, but as yet I wasn't sure why.

We had one more conversation in the defense ministry, with

Amos Lapidot, the air force commander. A taciturn fellow whose English came to him with some difficulty, Lapidot certainly had no difficulty singing the praises of the Lavi, which at the time I attributed to his official position and fear of political retribution. I thought, however, that he secretly was a Lavi opponent; I had heard rumors that several of the senior air force staff disliked the project. But Lapidot evinced nothing to that effect, not then nor afterward. My musings about him were due to the fact that I was still very new to the game of Israeli defense politics and had not done enough homework before making the trip. Had I known what I later learned, that then Brigadier General Amos Lapidot had led a Lavi briefing team to the United States in November–December 1982, I would have borne no illusions about his views. There was no way that the Israelis would have entrusted such a critical briefing to any officer who was not a true believer; Lapidot's laconic manner simply masked an enthusiasm that mirrored the outsized model of the Lavi that stood by his desk and dominated his office.

From the defense ministry we traveled to the Israel Aircraft Industries headquarters, near Ben Gurion Airport. Although we were treated with the utmost respect by David Ivri, the air force commander who had become chairman of IAI, and the rest of the senior staff, there was no doubt that we were in hostile territory. They did not like the briefing at all. Their top Lavi program manager and vice president for engineering, Moshe Blumkine, a large man with an overbearing mein—he was aptly nicknamed "Bully" by the Israelis, most of whom, I am in no doubt, were aware of the name's English meaning—seemed to be on the verge of bursting a vein in his neck. I was later given to understand that Bully always acted this way when he was upset; but then don't bullies always do so?

IAI recognized the threat we posed. Only a month earlier, a major feature in the trade journal *Aviation Week & Space Technology* had highlighted the importance of continued American financial support to the survival of the program.[3] The president of IAI was quoted as acknowledging that Israel could not afford to build more than half the plane in Israel, while Arens voiced his expectation that American financial assistance was needed not only for development but for production, including production in Israel. To Blumkine, the Lavi was a "state-of-the-art" aircraft, which nevertheless involved no great technical risks.[4] Small wonder that he disliked our nosing around his program.

Even IAI could not spoil lunch for me. I delighted in the delicious selections of meats that we were served; for once they were kosher and I could eat anything I wanted—within reason, of course. Normally, my luncheon fare consisted, at most, of fish. Many Orthodox Jews will refrain from eating fish too; some exceedingly strict Orthodox Jews will not even eat a salad not prepared under rabbinical supervision. For my part, I had compromised on salads, and such fish as met the Biblical standards of possessing fins and scales. But meat was a delicacy; I had never eaten a kosher meat meal while on business at the Pentagon. Even the IAI officials could not refrain from smiling when they saw how much I was enjoying their hospitality.

Our next stop was the finance ministry, which, like all Israeli government ministries except defense, is headquartered in Jerusalem. Normally, a trip from Lod to Jerusalem is uneventful. But ours was by helicopter, and the pilot took us around the Temple Mount. No sight is more moving for a believing Jew than the sight of the Temple Mount, the spiritual and emotional heart of Judaism for two and a half millennia. Yet to see it from several hundred feet in the air is incomparably more thrilling than to behold it from the ground. There is no obstruction to one's view; indeed, aircraft cannot fly over the Temple area because the airspace, like the ground, is holy. It is for that reason, incidentally, that Orthodox Jews will not venture up to the Dome of the Rock. For them it is sanctified terrain forbidden to all but the high priest, whose job has been vacant for a while.

It was a clear day; the golden dome on the Mosque of Omar (or Dome of the Rock), and the silver dome on the al-Aqsa Mosque gleamed as if they had just been polished. I closed my eyes, and opened them again, just to be sure I really was seeing what I was seeing. I stealthily put on my yarmulke, and uttered, in Hebrew, the catchall formula by means of which Jews thank the Almighty for special gifts: Blessed art Thou, O Lord our God, King of the Universe, who hast kept us in life, and hast preserved us, and enabled us to reach this festive season. The Lord must surely have been in a good mood: what seemed like seconds after my reciting the prayer, the pilot took us for a second circuit around the holy mount, careful as ever to skirt its airspace.

We landed at the Knesset's heliport, with the Menorah in the park opposite the parliamentary building in full view. From there we crossed the road to the finance ministry, to brief the senior staff. The

Israeli ministry of finance, like our own Office of Management and Budget and the British exchequer, and, indeed, all finance ministries the world over, is by tradition exceedingly skeptical about any activity that threatens to draw off disproportionate budgetary resources, especially from those of other agencies. Thus it was not at all surprising that our audience was especially uneasy about the Lavi program. In fact, one of their former colleagues at the finance ministry, Amnon Neubach, had conducted some rather extensive research into the Lavi's shortcomings. Since Neubach currently held the post of economic adviser to the prime minister, his opposition was of considerable importance, and he subsequently proved to be a vital ally as the Lavi issue took on a major public profile. I was to meet him for the first time later that day.

The finance ministry was the one friendly environment we encountered in Israel—other than my private off-the-record meetings with Neubach. The ministry team were extremely sympathetic to our frustration, but cautioned that we really did require Israeli cost figures if our critique was to prove credible. They also offered to arrange for us to meet with the finance minister, Yitzchak Modai, the following afternoon.

Under normal circumstances, a visiting U.S. government team would have taken the rest of the day off. We had already given three full-dress briefings, which was enough to drain off any excess adrenaline we might have possessed. But Fred Iklé was no ordinary government official, and the Israelis were no ordinary host government. Fred Iklé believed in working himself, and his staff, to the bone. A working trip was at best a busman's holiday. As for the Israelis, their standard operating procedure was to run their high-level guests ragged. Between the two, Israelis and Iklé, we had hardly a moment to breathe.

That evening, back in Tel Aviv, Defense Minister Rabin hosted a reception for our delegation. Minister Without Portfolio Moshe Arens was there, a man of middling height, with heavy-rimmed glasses and a monotonous voice that gave him a professorial mien. He looked as if I had punched him in the stomach when I introduced myself to him. The finance minister, Yitzchak Modai, was also there. The proverbial tall, dark, and handsome fellow, exceedingly well dressed, Modai had married a former beauty queen. He seemed a lively sort, quite the contrast to Arens; he agreed to see our group the next day. I also saw David Ivri, the IAI chief, and met Amnon Neubach, economic adviser to Prime Minister Peres, to whom I took an instant liking. Neubach is a very polished urbane sort; indeed, like other youthful officials associated with Shimon Peres, he always appeared so well dressed that his very image aroused the wrath of his rough-and-ready right-wing political detractors.

The top military brass were also there, including the air force and navy commanders. But in many ways the most important military officer I met was one to whom Amnon Neubach introduced me, Avihu Bin Nun, then the chief air force planner, who later would become the air force commander. A handsome man with piercing eyes and wavy hair that was turning silver, Bin Nun was the image of the dashing fighter pilot. All that he needed was the silk scarf and goggles.

As Bin Nun made very clear to me in the few moments we spoke, here was one pilot who most definitely did not like the Lavi. He worried that the program would drain resources away from higher-priority air force needs, such as acquiring additional F-16s. I knew that Bin Nun was by no means the only senior military officer who had doubts about the Lavi, even if his boss, the air force commander, was not one of them. A leading journalist, Eitan Haber, had reported nearly six months earlier that many senior Israel Defense Forces officers opposed the plane, especially senior army officers, who recognized that they would have to bear the brunt of the plane's budgetary costs. (Interestingly, in the same article, Haber had predicted that Rabin might reconsider his support for the program if it were found that development costs continued to rise. Haber ought to have known; he was sufficiently close to Rabin to later become his personal assistant at the defense ministry and then, in 1992, in the office of the prime minister.)[5]

But General Bin Nun was not an army officer. As a senior air force leader, his opposition connoted a particularly virulent brand of heresy that Arens was certain not to tolerate. Muzzled by his civilian and military superiors, Bin Nun knew that he was risking his career by talking to me. Nevertheless, he not only voiced his concerns about the plane but encouraged me to persist with my investigations.

Amnon Neubach later told me to protect Bin Nun. I needed no such instructions. My conversation with the general confirmed for me what I had already suspected: that canceling the Lavi was as much in Israel's interest as in America's. And I was not about to destroy the career of a man who clearly displayed so much political guts.

I had one other encounter at the reception, which, by the way, was kosher, of course—I never could get over the delight of eating meat at every function I attended, and they all seemed to have meat dishes. As I was standing at the smorgasbord, a very attractive young woman approached me. She had, I recall, jet-black hair and very dark eyes. She had sought me out in the crowd, and she proceeded to tell me that the defense minister, Yitzchak Rabin, secretly supported what I was doing, that he opposed the Lavi. She quickly left my side, and no one I spoke to thereafter knew who she was. I never saw her again, and do not know her name. Our conversation mystified me at the time, and continues to do so to this day.

Monday, our second full business day in Israel, was every bit as hectic as the first. We drove to Jerusalem in the morning, with our first stop the prime minister's office. Shimon Peres had been on the Israeli political scene as long as his archnemesis and Labor Party colleague Yitzchak Rabin. Like Rabin, he was a boy wonder, serving as the director general of the defense ministry when he was but thirty, and when the defense minister was none other than the indomitable David Ben Gurion. Unlike Rabin, Peres had never before served as prime minister—he had been Rabin's defense minister when the Entebbe operation took place and I was a tourist at a Jerusalem hotel. Peres had become prime minister as part of a deal he had made with the Likud leader Yitzchak Shamir in creating a National Unity government of the Likud and Labor parties. Peres was a famed dealmaker—it was said that was why the Israeli public didn't really trust him—and the National Unity government looked like the best deal he had made. He was to serve for two years as prime minister, to be followed by Shamir, who would also serve two years.

Clearly, one needed a scorecard when dealing with so many men who either were, would be, had been, or wanted to be prime minister. As in most parliamentary democracies, top Israeli politicians floated from ministerial portfolio to ministerial portfolio, always, however, with an eye on the ultimate prize. But Israel was different in that within both of its major parties no one could ever be sure of staying on top for long, nor between those parties was it clear who would emerge the ultimate victor. The electoral gap between Labor and Likud continued to shrink while the power of the smaller parties to throw their weight behind one or the other of the larger ones continued to grow. And that was all without a National Unity government—indeed, that was why one was created. No doubt these political factors further complicated the positions of the top policymakers on a highly emotive issue such as the Lavi; few could afford to alienate large segments of their own party, much less the electorate as a whole.

In some respects Yitzchak Shamir, the least telegenic and charismatic among the ruling elite, was the shrewdest of them all. It certainly seemed that in agreeing to the National Unity government, the wily Shamir outfoxed his counterpart, Labor leader Shimon Peres. Peres ably oversaw the withdrawal of Israeli troops from Lebanon and pulled the Israeli economy back from the brink of hyperinflation during his two years as prime minister from 1984 to 1986. Nevertheless, it was Shamir, his successor as prime minister under the National Unity agreement, who was able to exploit his incumbency to win reelection to the premiership in 1988. Then, in the renewed National Unity coalition government, Shamir exiled Peres to the politically uncomfortable job of finance minister, for which he had little aptitude and less interest.

But that was all in the future. In April 1985, Peres was in charge, having masterminded the retreat from Lebanon and initiated the economic revitalization of his country. Like most people who have met him, I found Peres to be an impressive, even striking figure. His well-tailored suits, his high combed-back hair, and his almost regal bearing made him *look* like a prime minister. His intelligence made him sound like one too. At the same time, he had the natural politician's knack of coming across in a friendly manner and putting his interlocutor at ease. It wasn't hard to brief Shimon Peres.

Peres stressed that he was a strong supporter of high-tech industry, and, as a natural consequence, supported the Lavi. He obviously had not been listening to Neubach, as I later told Amnon, who simply shrugged his shoulders and said, "That's politics." Nevertheless, Peres did not close the door completely on the idea of altering the program. He listened politely, his "no" was not a "hell no," and for me, that was an improvement over my reception at IAI the day before.

Our next stop was a return visit to the finance ministry, to meet with Yitzchak Modai. Modai had at best a mixed reputation in Israel. He was leading figure in the Liberal Party, a small free-market, highly secular, and somewhat hawkish grouping that had merged with Menachem Begin's right-wing Herut Party to form the Likud. A successful businessman, Modai was known for his quicksilver temper and his uncanny ability to say the most inappropriate things at the least opportune moments. But he is a very bright man, and had been performing quite admirably as finance minister, working closely with Peres to rescue Israel from the specter of hyperinflation that haunted the economy.

Modai's hawkishness was outweighed by his fiscal prudence, and we considered him a potential recruit on the Lavi issue. Modai listened to our briefing most attentively, and clearly absorbed our message; he welcomed our efforts and offered his support. Modai emphasized that the decision lay primarily in the hands of the defense ministry; nevertheless, his status as a member of the National Unity government's inner cabinet made his an important voice on the fate of the plane. We hoped that, consistent with his reputation for bluntness, he would indeed speak out when the time came to do so.

We were scheduled to meet with Defense Minister Rabin again later in the day, but before we left for that meeting Iklé had arranged a private tête-à-tête with Minister Without Portfolio Moshe Arens. Normally, I, like all assistants, deputies, and other miscellaneous horse holders, would have been upset at being excluded from a meeting between my principal and the minister of another country. But after the chilly reception Arens had given me the night before, I had no great desire to see him. Instead, Scott Redd, Phil Merrill, and I wandered around the Old City while Fred Iklé paid his courtesies to Moshe Arens. This brief interlude was another high point of my trip and probably the highest point for Scott. A devout Christian who keeps a large Bible on his desk at work—it is opened to the chapter of Proverbs, which has thirty-one chapters, that corresponds to the day of the month—Scott had never visited Jerusalem before. He availed himself of the time Fred gave us to walk the Stations of the Cross and to soak up the religious atmosphere that governed his life. Not much later, Scott, then a navy captain, contracted a particularly nasty form of cancer. If that were not bad enough, our office was haunted by the eerie fact that one of his predecessors as Fred Iklé's military assistant, like Scott a brilliant young navy systems analyst, had died of cancer only a few years before. But Scott was first and foremost a man of faith, the embodiment of what people have in mind when they think of a good Christian. His faith stood by him just as he had long stood by it. Scott recovered, though, as he told me around the time that recovery became a distinct possibility, he never expected to serve at sea again. But Scott did put to sea, serving his country in the North Atlantic, the Mediterranean, and in the Persian Gulf, as commander of the navy's Fifth Fleet. He wears three stars on his shoulderboards. I don't know who is more fortunate, Admiral John Scott Redd, for having been given the chance to serve his country again, or his country, for having received so much from such a magnificent individual.

While Scott was soaking up the legacy of Christianity, I repaired to the holiest site of my religion, the Western Wall. Popularly called the Wailing Wall, it is the outer wall of Herod's magnificent reconstructed Temple Mount, and is the only remaining artifact of the original Temple structure. I had visited the Wall before, but each visit, particularly since my very first as a teenager, has had for me its own special meaning, if not magic. I was to visit the Wall many times during my work on the Lavi project; this time, I prayed to the Lord to forgive my sins—of which I have never been short—and gave *tzedakah*, charity, to the many outstretched palms that accosted me as I crossed the courtyard from the ritual washbasin to the Wall.

We all regrouped shortly thereafter for a final meeting with Rabin. The defense minister was as genial as he ever was, which is to say he didn't frown very much, if at all, and even offered a hint of a smile. He said that he appreciated the work we had done, and was prepared to cooperate with us on this important matter. He didn't encourage us, but didn't discourage us either. Iklé offered to deliver a report to him on where the costs really stood, and Rabin agreed. That was the opening I needed. I boarded the plane for Pakistan (via Egypt) feeling that the first part of my mission had been completed successfully.

4

Meetings, Meetings—in Washington, in Israel

W e had fired our opening volley, and the Israelis had retreated. On the orders of the defense minister, they were now ready to respond to our requests for detailed information on the Lavi program. At Iklé's request, I put together a schedule for following up on the trip and producing a report to Rabin. I estimated that the effort would actually consist of two phases. The first would involve the cost evaluation that we had announced we wished to undertake. This phase was to require some eight to twelve weeks of research and analysis. After another meeting with the Israelis, we would initiate a second phase, although I recognized that it might overlap with the first. This second phase would focus on alternatives to the Lavi. It would extend for nearly two months and culminate in a negotiation with the Israelis over both the alternatives to the Lavi and related industrial "offsets" in order to sweeten the bitterness of abandoning the program. These offsets would encompass work undertaken in Israel as subcontracts to American defense firms, in order to help generate revenues to pay for the proposed Israeli arms purchases—in effect a sophisticated form of barter.

I also established an interagency Lavi Steering Group, with senior representatives from State, the National Security Council, and the Office of Management and Budget, as well as from the air force and the Office of the Secretary of Defense. My objective was to ensure that our analysis would not be simply the product of the Pentagon, which had been vilified in the Israeli press ever since Caspar Weinberger took office. The steering group was nominally at the "two-star" level, that is, major generals or their civilian equivalents. In practice, attendance, except for the more important meetings, was mostly by "the colonels," senior aides to the nominal members whose business it was to keep the wheels of government running smoothly.

Our first task was to formulate questions to put to the Israelis. Working with the team that had produced the briefings, especially with the air force's Linda Hardy, we compiled four pages of detailed questions covering every aspect of the program. We first outlined some of the basic data we needed to make our estimates. Such data included the airframe's weight; all details about the engine; materials used in the airframe; the plant layout; the make/buy plan (that is, how much the Israelis expected to produce and how much they would acquire from others); performance requirements; current cost estimates (as opposed to the outdated material we had been given); and the methodology for the Israeli estimates.

We also sought information about the way in which the Israelis hoped to support the plane. We asked for Israel's reliability and maintainability goals, as well as its maintenance concept. We asked whether such goals and concepts were considered in design trade-offs. Answers to these questions would shed light on the cost of operations and maintenance over the life of the plane, and therefore on what are termed "life cycle costs." We asked if there was a plan for spare parts, and requested information about the status both of the design and of work on the transition to production. We also asked for details on the program's automatic test equipment.

We had questions about the program's approach to testing. In many, if not most, cases, manufacturers sloughed off the demands of testing, often advertising test schedules that were far too optimistic (that is, too compressed) ever to be met realistically. Delays in test schedules and disruptions to those schedules caused by unbridled optimism invariably increased development costs. We suspected that the Israelis, in their zeal to minimize costs, had concocted an unrealistic test schedule, and we therefore requested facts about the testing "philosophy" and its associated test plan. We sought information regarding the estimates the Israelis had made for the costs of engineering changes resulting from thorough testing—we suspected that such estimates did not exist. We also sought the details of the expected impact of testing on development and production; again, we suspected that little would be forthcoming.

We asked about the plane's operational missions. Its four missions—close air support, air defense, training, and special missions—required varying performance characteristics. We suspected that, as is often the case, optimizing the plane for the performance of one mission came at the expense of performance in the others, but we needed detailed material to validate our hunches.

Other questions addressed the training program; anticipated and realized development delays; production schedules for major subsystems and components; and, critically, the work breakdown structure (WBS) and the associated WBS dictionary, which highlighted each portion of the overall effort. The WBS actually provided the best tool for cost estimation, since by outlining just how the work was organized, it enabled analysts to estimate each activity or sub-effort, to be aggregated thereafter into a full and final estimate. Because the WBS held the key to the Lavi's true costs, we anticipated its reception eagerly; Linda Hardy in particular could not wait to get her hands on this jewel.

Needless to say, the Israelis were less than thrilled by our request. Their response was to propose that we meet with a delegation of their specialists in Washington later in June. This, of course, delayed the exercise, but we had no alternative but to welcome our prospective guests when they were prepared to visit us.

It was instructive—and not a little bit annoying—that while we had to wait for program manager Menachem Eini and company until June, a Lavi team consisting of IAI and Israeli air force officials was making the rounds of leading American aircraft manufacturers during the third week of May. The Israelis were unabashed about the purpose of their visit. As one of the briefing papers put it, they were in the States "to learn how GD [General Dynamics], McDonnell, and Northrop integrate avionics systems." System integration for the Lavi was scheduled to begin in late 1986. At GD, for example, the Israelis were given a presentation on F-16 system integration, software configuration control, and laboratory systems. They were also treated to a laboratory tour.

The Israelis did not pay any of the American manufacturers for the privilege of picking their brains. That the Israelis undertook this exercise underscored to many of us in the Pentagon just how dependent they were on American know-how, even in areas for which they were supposedly not relying on American contractors. The contractors, for their part, knew exactly what was going on, but could do little about it. McDonnell and GD had both recently sold aircraft to Israel; Northrop still hoped to do so. Not one of them was prepared to alienate so important a potential customer, especially since it was widely recognized that any sale to Israel vastly enhanced the prospects of making additional sales of the same system to her Arab neighbors.

Although I was aware of the Israelis' visit, during the weeks before the Israelis arrived in Washington I did not have much time to think about them or the Lavi. I was preoccupied with managing the *Defense Guidance,* whose publication was nominally due in October. Indeed, meetings I held on the Lavi were often immediately followed by meetings on the *Guidance;* I simply could not concentrate on the Lavi as much as I would have liked, and therefore generally let my team proceed according to its own best lights. I also needed a rest, and I was able to get away from Washington for a few days of vacation without the kids.

Acting on little more than a whim, we booked ourselves into Grossinger's, the doyenne of the Catskills. I had never stayed at any of the really top-flight Catskills resorts before. My parents had rented a small, one-bedroom bungalow in nearby South Fallsburg for several summers, and I had visited some of the second-tier hotels on occasion. But never the big ones: Brown's, Kutscher's, the Concord, or Grossinger's.

To say that we were disappointed in Grossinger's is an understatement. The facility looked like it needed a face-lift. I wasn't interested in competing with octogenarians for the "Simon Says" championship. The weather was overcast, so I couldn't even sun myself at the poolside. I knew none of the other guests.

What Grossinger's had was food. Lots of food. Not particularly great food. But lots of food. Entrées galore. Order as many as you want. You don't like what you got? Ask for another entrée. Or eat them both. Or order all six. And don't forget the side dishes. And here's more rolls. Everybody loves the rolls. Pickles too. You want a special? No problem. Specials are our specialty.

We left after two days, bloated and bored. Grossinger's shut its doors a season or so later, and I understand that it has been acquired by a Korean group that hopes to cater to the influx of Asian-Americans into the Catskill region. Good luck to them. It was a pleasure to be back at the Pentagon. No specials there.

I gathered my team together on the morning of June 10 to review the bidding before the Israelis arrived later that day. Armed with the knowledge that Defense Minister Rabin supported our request for data, we were determined not to let his minions stonewall us, as we surmised they might. Linda Hardy was particularly eager to match wits with her Israeli counterparts.

We held our first meeting with the ministry of defense team that afternoon. Their group was headed by Zvi Tropp, economic adviser to the defense minister, and Menachem Eini, the MoD's Lavi program manager. Our specialists included the colonels and their equivalents from our steering group: among them not only Mike Foley, Linda Hardy, and some other air force personnel, but also representatives from the State Department's Office of Political Military Affairs, the Office of Management and Budget, and some of the PA&E personnel who had prepared our initial estimates, as well as Dick Smull of my office.

I enjoyed interacting with Zvi Tropp, a tall, genial, exceedingly bright fellow. In his capacity as chief economic guru of the defense ministry, he operated as a sort of director of program analysis and evaluation. Yet his staff consisted of all of six people. Moreover, half of them were devoted to assessing macroeconomic problems, such as the impact of the defense budget on the Israeli economy. Thus there were only three people to review and evaluate all of Israel's multitude of defense initiatives, developments, and procurements, both from within and outside the country. It was clear that Tropp did not exactly have the resources to give the Lavi program its full due.

Tropp's inability to focus on the Lavi left all matters in the hands of the mercurial Eini, who was certainly no joy to work with. The little Iraqi was no great analyst himself, however, but he seemed to be very thick with Bully Blumkine and his buddies at IAI, leaving us with the clear impression that it was they who really pulled the analytical strings.

Eini's manner from the outset of the meeting made it impossible for us to avoid the conclusion that he was unhappy to be in the States, unhappy to be disclosing information to the Americans, unhappy to be forced to justify his program. Moreover, he almost immediately made a massive strategic error. He took Linda Hardy head-on. Perhaps we should have expected that he would do so. Linda was young.

She was a soft-spoken woman. And she was black. Eini, reared in an Oriental Jewish environment where women made babies and cooked spicy foods, and where blacks, if they appeared on the scene at all, did so as servants of one sort or another, simply couldn't comprehend that he was dealing with one tough analyst. The rest of us knew better, and we sat back in amusement as the IAF combat veteran and the USAF staff specialist locked horns. Eini argued that the information we were seeking was unnecessary. Patiently, but firmly, Linda outlined why it was very necessary, and puzzled aloud as to how Israel could invest billions into a program without having at its disposal the facts that she was seeking. Eini visibly gnashed his teeth, and groused to his teammates in Hebrew, little realizing that I understood everything he was saying. (That scene was to be repeated a few more times, until some days later, I finally let on that I had caught every one of Eini's aspersions in the holy tongue.) Linda continued to hold her ground, winning the admiration not only of her American colleagues but also of the more professional of her Israeli interlocutors.

We did not meet with the Israelis the following day; they were studying our request for information. That night we hosted the delegation at my home. The Israelis all turned on the charm, and were treated to a kosher meal in return. Since I was at home, I wore my yarmulke; it was the first time that Eini and the others realized that I was an Orthodox Jew. The only exceptions on the Israeli side were Eli Rubenstein, the newly arrived minister at the Israeli embassy, and his wife. Eli, also an Orthodox Jew, had joined our synagogue, and I would usually chat with him after services; we had entertained him on Shabbat a few weeks earlier before his wife, Miriam, had arrived in the States. Eli has since moved on to greater things, serving as cabinet secretary under both Shamir and Rabin and as the lead Israeli negotiator both in the post-Madrid peace talks with the Palestinian-Jordanian delegation and in the successful treaty negotiation with Jordan.

Our next meeting was scheduled for 11:15 the following morning; an hour earlier I was able to catch part of a discourse on Israel and American Jews given by Rabbi Arthur Hertzberg at the American Enterprise Institute. I had never seen Hertzberg before, though I knew quite a bit about him. A genial and highly articulate Conservative rabbi and an academic of some repute, Hertzberg was also an unabashed liberal who applied his ideology to Israel's circumstances as well as to those at home. Accordingly, he was a strong advocate of peaceful resolution of the Arab-Israeli conflict through accommodation with the Palestinians. It was a position that earned him the undisguised contempt of the American Jewish leadership, a contempt that he reciprocated in full.

Hertzberg asserted that the Jewish establishment slavishly toed Jerusalem's line. He felt that in fact the American Jewish leadership was—to mix metaphors—holier than the pope, since many Israelis, perhaps the majority, disagreed with the government's hard-line stance vis-à-vis the Palestinians. I was not in much agreement with Hertzberg's views about the Palestinians, but I took his remarks about the Jewish leadership as a warning for me personally. If Arthur was correct, and I had no reason to believe he was not, I stood to encounter strong opposition from the American Jewish leadership even if Rabin were to announce that he supported our views. The defense minister had as yet made no such announcement, and, other than his willingness to entertain our analysis, and the encouragement of a mysterious individual who claimed to know his thinking, I had no reason to believe he ever would.

Hertzberg's presentation was completed just before I was due to meet again with the Israelis, and without having had a chance to speak to him personally (as I would many times afterward), I rushed back to the Pentagon.

The tenor of talks that day was no better than that of two days earlier. The Israelis essentially said they would see what they could come up with, denying in particular that they possessed a work breakdown structure. Linda Hardy was highly skeptical; she could not see how anyone could proceed without a WBS. The Israelis claimed that somehow they did their business differently. We, for our part, were convinced that while wage rates might differ in our two countries, building a sophisticated fighter involved the same work everywhere. Our experience with our NATO allies confirmed that view.

We encountered similar problems as we sought to get some clarification on issues relating to reliability and maintainability (r&m) and to logistics. The Israelis essentially stonewalled us, although I insisted to them that r&m was the basis for readiness and operations and maintenance cost estimates. The fact that our own r&m specialist, Marty Meth, had not heard from the Israelis whether he even had a counterpart in the IAF was not a very good sign at all.[1]

It was clear that we were not going to get much more out of the

Israelis on this visit, and it was agreed that we would make a fact-finding trip to Israel later in the summer both to receive detailed answers to our questions and to follow up with officials at both the defense ministry and IAI.

Matters other than the Lavi preoccupied me for much of the rest of June and July. There was relatively little public comment on the issue during that period, although *Aviation Week* carried a lengthy letter in its June 10 issue from a gentleman who went into great detail to demonstrate that "stupidity and politics, in both the U.S. Congress and the Israeli Knesset, have again prevailed over reason as the Israelis proceed at full speed to reinvent the wheel at the expense of the U.S. taxpayer and the federal deficit." He concluded by asserting that "the Lavi fighter program is a waste of money and should be stopped immediately. Israel's defense and the U.S. taxpayer's pockets can be better served by providing Israel with additional U.S. fighters, not inferior copycat versions of U.S. fighters."[2]

I had no idea who the author of the letter was or what his affiliation might have been. But I knew that he was right on target. Unfortunately, he also represented Moshe Arens and IAI's worst nightmare. For it was exactly the kind of sentiments that the letter expressed that could spell the doom of their program. In any event, we were nowhere near the point where we could suggest alternatives to the Lavi. That was for a future time. Maybe.

There was one other matter that I had to deal with before we departed again for Israel. On July 9, Mel Levine, a Democratic congressman from the Los Angeles area, wrote to Cap Weinberger requesting "the Lavi report in your possession pertaining to the recent consultation on the matter between our government and the government of Israel."[3] Levine, a personable type, looked at least ten years younger than his real age. I once bet him a case of wine that he was younger than I was—and lost; it wasn't even close! Levine was very much pro-Israel, but no knee-jerk supporter of the Israeli line. The Lavi case was especially difficult for him to deal with, because his district included many aerospace workers (especially at Northrop) who were not sympathetic to the idea of taxpayer support for foreign competition to their products.

Levine formulated his request in the context of an amendment to the foreign aid authorization bill that dealt with Lavi funding; he was a member of the House Foreign Affairs Committee, which had oversight responsibility for the foreign aid budget. It was unclear to us whether he actually knew that we had issued no report and that the only existing documentation was the copies of my April briefing to the Israelis. We did know that the Israelis preferred that the briefing not circulate: it did not paint the Lavi in a particularly good light. It was therefore not very hard to parry the congressman's request; we simply told him that no report had as yet been issued.

My next task was to finalize the dates for our upcoming visit to Israel. I decided to schedule the team's departure for its next visit to Israel on July 29, the day after the fast day of Tisha b'Av, the ninth day of the month of Av, which is the most miserable day on the Jewish calendar. The fast commemorates the burning of both Temples; the rabbis of the Talmud, not content to assign only these two tragedies to the fast day, piled on others as well, so that it has come to encapsulate all that has been calamitous in Jewish history. Interestingly, World War I began on Tisha b'Av; some rabbis have argued that since World War II and the Holocaust were both due to Germany's defeat in World War I, they too could be linked to the ongoing Jewish tragedy that the fast commemorates.

I have never liked Tisha b'Av much. To begin with, who likes a fast day, except perhaps people who are overweight? I'm not overweight, or at least I wasn't at that time, and fasting gave me, and still gives me, a severe headache. Secondly, while the rabbis have also taught us that someday Tisha b'Av will be transformed into a holiday, it is very unpleasant to go hungry in the heat of the summer. Tisha b'Av, like Yom Kippur, also imposes other uncomfortable burdens on Jews: no shoes; no washing; no shaving. Not pleasant in the summer at all. I'm sure it's easier for my brethren in the southern hemisphere.

Because tradition has it that Jews should undertake no stressful activities during the eight days that precede the fast, as well as, of course, during the fast itself, I even elected to schedule our pretrip meeting prior to the commencement of the month of Av. I did not expect to have a marvelous time in Israel; the last thing I needed was to tamper with hoary tradition as well. Was I being superstitious? You bet I was!

Although we had already given the Israelis a list of detailed questions, they asked us to spell out specifically just what it was that we desired when we came to Israel. After all, in their view, they had already revealed far more than was necessary for us to give our blessing to the program. We recognized what really lay behind their request: the Israelis were trying to ensure that they could "game" our visit by providing specific answers to some, if not most, of our questions, without inadvertently blurting out information that they would rather we didn't have.

We were not sure what exactly such "information" might be. But we were certain that it was not merely a matter of the Israelis' viewing our demands as a nuisance, which no doubt they did as well. The project was so large, so expensive, potentially so laden with controversy, that any indication that Israeli estimates were based on something less than rigorous analysis could be extremely damaging to its fortunes, and to American-Israeli relations as well.

Moreover, the possibility of gross underestimation of program cost was hardly a remote one. It was precisely because rosy cost estimates were a common failing of military bureaucracies everywhere that we sensed that the Lavi was a program in trouble. The Israelis contended that they were somehow different, that their costs could be controlled more rigorously than those of analogous programs in other countries. Their credibility was therefore at stake, and they had little sympathy for our view that, given the billions we were pouring into the plane, we saw no alternative but to get the facts we needed.

Accordingly, my team once again drafted a second set of detailed questions that had to be answered if we were to generate accurate estimates of the Lavi program. But our people really were acting in good faith; they did their best to minimize the obvious burden on the small Israeli staff. For example, one official indicated that in forwarding his reliability and maintainability questions to the Israelis he presented them with a choice for estimating reliability using either a more objective parametric approach or an approach involving engineering judgment, which, of course, was far more subjective. He even offered to run the parametric estimate on his own computers.[4]

On July 9, 1985, I transmitted forty detailed questions to the Israeli air force attaché in Washington. I indicated which of those questions could be answered on paper and which required face-to-face discussion.

Our follow-up questions fell into several different categories. The first category was of the catchall, general variety. It included a request for information on the design and production cost history of the Kfir aircraft, the plane the Israelis had adopted from the French

Mirage III design. This history would give us a better sense of any cost peculiarities that were due to the nature of the Israelis' design and production system. We also sought a forecast of their business base and overhead rate, so that we could better estimate both program and individual unit aircraft costs as well as obtain some insight into their overhead costs.

We also inquired about their production process, so that we could understand how it differed from that of the United States. We needed to view their plants and examine their machinery. We needed some insight into their employment policies, to see how they differed from ours and what the cost impact of those differences might be. We also needed estimates for the Lavi's planned design and test hours.

The remaining categories of questions dealt with detailed, technical issues. We asked for estimates of the cost of developing and procuring all support elements, and for the rationale underlying the generation of these estimates. We wanted the Israelis to specify their various reliability and maintainability goals—that is, the degree to which the plane could be expected to operate without major failure and the amount of time spent on maintaining the plane relative to its operational use. By outlining their goals, the Israelis would facilitate our estimation of spare-part and support requirements (and therefore costs) over the life of the program. We also wished to compare such factors with those for the Israeli F-16, to see whether Israel had developed special cost-saving techniques that justified lower estimates than those generated by our own cost models.

We felt that we needed to discuss in detail maintenance concepts, such as for the avionics, built-in testing, and composite materials, and the cost implications of these concepts. My team again requested the work breakdown structure, including that for the development of support equipment and technical data.

Lastly, we informed the Israelis that we needed to discuss their integrated logistics system plan. Such a plan would indicate how they developed and hoped to apply their overall approach to the use of spares, depots, and test equipment. By understanding the plan, we would also obtain better insights into their projected program costs.

I had no doubt that our questions would cause considerable consternation on the part of Eini and his program office. If they gave us complete answers, we might be able to demonstrate that their costs were far higher than they had postulated. If they refused to be responsive, we could charge that they were flouting the defense minister's instructions. And if they claimed that they didn't have the data at all, we could question the credibility of their management, as well as of their estimates.

The technical team that I took with me to Israel for a weeklong study visit beginning July 29, 1985, found the Israelis forthcoming on some issues, but not on others. The team, which included members of our interagency group, as well as the air force's Colonel Mike Foley and Captain Linda Hardy, did ascertain that the Israelis were having serious problems with major elements of their program. The Lavi appeared to have stability problems because of tail-heaviness. The Lear Siegler flight control system, which was being installed on both the Lavi and the Swedish Gripen, suffered from malfunctions in both aircraft. The Lavi was also experiencing system integration difficulties. Worst of all, the plan for coproducing the plane's Pratt & Whitney PW1120 engine had completely collapsed.

I had already learned of the engine program's problems in an article that appeared in the *Hartford Courant* on July 2 and was reproduced in the DoD "Yellow Bird" the following morning. The "Yellow Bird," formally called "Current News," is the daily set of clippings that the department's public affairs office compiles from that day's *Washington Post, New York Times, Wall Street Journal,* and *Washington Times,* as well as from the previous day's editions of other major and not-so-major newspapers. The *Courant* fits into the latter category; it is important for defense readers because it tends to have the best-informed coverage of the Connecticut contractors, for example, the Electric Boat shipyard and United Technologies and its Pratt & Whitney engine subsidiary.

The *Courant* reported that the plan to coproduce the PW1120 engine for the Lavi was likely to be canceled. The Israeli Bet Shemesh engine plant, nominally the "prime" contractor (with P&W as its subcontractor), was having such severe difficulties that production schedules were reportedly three years behind schedule.[5]

The situation was actually worse than even the *Courant* had reported, and it offered some real insights into the management problems that were bedeviling the aircraft. The Israelis had planned initially to coproduce the engine with Pratt & Whitney, and then to produce the follow-on engines entirely on their own. In the event, the Bet Shemesh engine plant was incapable of carrying out even the initial, more limited, task. Pratt & Whitney had reapportioned the coproduction work several times, giving the Israelis increasingly less complex tasks. Finally, frequent changes of management, labor problems, and other management deficiencies forced the cancellation of the coproduction effort only a few weeks after our visit to Israel in April, although the decision to cancel coproduction remained a closely held secret.

Naturally enough, the MoD program office and the IAI people had little to say to my team about the engine when we began our round of meetings at the ministry in Tel Aviv on July 30. Elsewhere in the ministry, however, some individuals were a bit more talkative. We learned that the Bet Shemesh fiasco not only was a blow to the senior Lavi project leadership but also had imposed a rather large additional financial burden on the government, which had to bail out the plant. It was difficult to estimate just how much that bailout was likely to total. The *Courant* had suggested a sum of $40 million, but that was prior to complete cancellation. In any event, I decided not to factor the cancellation costs into the cost of the Lavi. In fact, I did not account for any costs associated with what analysts termed "schedule risk"—that is, the price of delays. I reasoned that my estimates should be viewed as conservative, at least by those without a partisan interest in the matter; rising costs because of schedule lags would simply underscore the validity of what we had done.

The program office's reticence regarding the engines extended to other data that we requested. In particular, we ran up against a stone wall in our search for the work breakdown structure, or any information that could shed light on it. Perhaps the Eini team's readiness to defy some of our requests may have been due to their perception that Rabin, while willing to humor Weinberger and the DoD, had no serious intention of curtailing the program. They certainly had reason to take heart from his response to a very blunt question in a July 19 Israeli radio interview (that is, just ten days before our arrival in Israel):

"You are now being quoted," ran the question,

> as having said that if we could go back to the past, you would reconsider the Lavi project, but now there is no going back. However, in view of what is occurring in the economy, in view of what might yet occur, why is there no going back on . . . a project which . . . could perhaps save the Israeli economy?

Rabin labeled the questioner's assertion "baseless." He argued that it would cost Israel just as much to purchase aircraft from the United States as to build the Lavi. He added:

If we do not develop the Lavi . . . and manufacture it . . . we will . . . have to devote the same sums to purchasing planes from the United States, because Israel is investing almost none of its own resources in the Lavi's development, and this applies to its manufacture as well. . . . U.S. money . . . is being given to be used either for the development and manufacture of the Lavi in Israel or for the purchase of a suitable equivalent air force plane from the United States.[6]

While Rabin seemed to be taking a hard-nosed stand on the issue, I was reasonably sure that he was in fact posturing. Much as in 1958 Charles de Gaulle had told the Pied Noirs who supported him and thought he backed a French Algeria, *"Je vous ai compris,"* Rabin was being very careful about his commitment to the program. The key was his statement that American money was being used either for Lavi or for an American plane. That was exactly our position, except that by now we were reasonably sure that, once all the facts were in, the money would indeed be spent on an American plane.

Our visit to Israel was the first of many that I was to lead over the next two years in the course of our efforts to learn more about the Lavi. Because it was a weeklong affair, it afforded me my first opportunity to work in depth with some of the recently arrived members of the U.S. embassy team. It was a pleasure to make a friend of Tom Pickering, the new ambassador, a professional's professional if ever there was one. Tom knew the Middle East well, having served as ambassador to Jordan some years earlier. He had just completed a very difficult assignment as our ambassador to El Salvador, where he had assisted the beleaguered government in its efforts to quash a prolonged insurrection by leftist guerrillas. Tom, who has since moved on to serve as U.S. ambassador to the United Nations, India, and Russia, cut a very different figure in Tel Aviv from his predecessor, Sam Lewis. He did not seek, or receive, the sort of adulation that showered Sam everywhere he went in the country. His relations with the Israeli establishment were more formal, more correct. Some Israelis mistook Tom's attitude for antipathy, which it certainly was not. He simply approached his job differently than his predecessor had, and in so doing served his country no less ably.

Tom was very sympathetic to my efforts to shed some light on the Lavi program. He showed a keen interest in the details of our analysis and offered very useful insights into how we might best make our case with those Israelis who would have the most to say about the plane's future. I was delighted that he was always ready to accompany me to meetings with Rabin and some of the other senior people; his presence would invariably convey to our interlocutors both the gravity of our concerns and the fact that they reflected the views of the U.S. government as a whole and not just of Caspar Weinberger and the Defense Department.

I was also exceedingly fortunate in receiving a great deal of support from our defense attaché's office. The embassy had a new senior attaché, Colonel Joe Bavaria, a very bright air force officer with a good sense of the nuances of Israeli politics in general and security politics in particular. Like Tom, he was always available with excellent insights and advice, from which we profited greatly.

Our daily point of contact, and an integral part of our team, was one of Bavaria's deputies, Steve Ham, assistant air force attaché, who, like any good attaché, served as our eyes and ears in Israel. Steve, a decorated fighter pilot, is a tall, wiry man with a sharp sense of humor and, like many flyboys, has the ability both to tell a good yarn and to comment graphically on the charms of Israeli women, of which there are many. He must have terrific peripheral vision: as he drove us around Tel Aviv he could evaluate countless Israeli female pedestrians without ever colliding with another vehicle. As for his stories, one of his better tales was about one of his predecessors who supposedly had been approached by the Israelis with photos of himself in rather compromising positions with a beautiful Israeli intelligence agent. His response was probably not what the Israelis expected: "Send me enlargements; my friends would never believe it!" If the man existed at all, he was probably an aviator.

Steve Ham had no great love for the Lavi; he was convinced it was a waste of everybody's money. He was continually providing us with information about press and public attitudes to the plane, which was invaluable as we mapped out our own strategy to keep the issue in the public eye. He also afforded us considerable on-the-spot analytical acumen, a welcome supplement to the considerable brainpower—no fewer than thirty hardy souls—that we had seconded to the effort back home.

Before returning to the States on August 4, I managed to spend the Shabbat in Jerusalem, staying at a very pleasant hotel called the Laromme, which was soon to be my favorite in Israel. The hotel is located near the picturesque Yemin Moshe neighborhood, named after Sir Moses Montefiore, the British financier and philanthropist; its most notable landmark is a windmill that Sir Moses erected in the mid-nineteenth century. Until 1967, Yemin Moshe was well within the range of Jordanian guns and was, as a consequence, virtually uninhabited. Now it consists of terraced homes and flats, ornamented with bougainvillea and other colorful plants and offering some of the best vistas of the Tower of David and those same city walls that had offered cover for sniper fire.

From the Laromme, the trek to the Western Wall takes about twenty minutes. One proceeds down steps that intersect Yemin Moshe and Mishkenot Sha'ananim. Yemin Moshe was the first Jewish neighborhood outside the Old City Walls. "Mishkenot," as it is popularly called, was built by Montefiore between 1855 and 1860. It now is the Jerusalem municipality's guesthouse for visiting artists and academics. At the bottom of the stone steps is the Valley of Hinnom, the Gehinnom—Hell—of Jewish religious lore, which now is the site of an outdoor theater. Crossing the valley and dodging the light Sabbath traffic—driving in Jerusalem's traffic snarls during the workweek is an exercise in pure masochism—one climbs a path named after an Israeli soldier named Benny, from which the view of West Jerusalem is breathtaking, and proceeds on a path alongside the massive Dormition Church and gardens through the Zion Gate and into the Jewish Quarter. By 1985, the process of restoring the quarter, which had been completely demolished by Jordanian forces in the aftermath of the 1948 War, had moved far along toward completion. The new buildings gleamed, and yeshivot, seminaries for advanced Talmudic study, were sprouting like mushrooms. Several yeshivot were being constructed around the perimeter of the Western Wall's plaza, so that the entire compact area was a beehive of religious activity.

On my way back from the "Kotel," as the Wall is called in Israel, I found myself walking through the Batei Machse Square, an area that had been built for poor Orthodox Jews in the late nineteenth century, destroyed after the 1948 War, and rebuilt since 1967. As I crossed the square, I saw some small boys and girls, dressed in Shabbat clothes, running around, playing some game. I must have wandered nearly across the open square, lost in thought, when I looked up and saw that I was walking directly toward two religious old people, an old man and old woman, she with her hair completely covered, sitting at the edge of the square. The sight of them caused me to stop short; my mind had flashed back Proust-like to a story at the tail end of the Talmud's Tractate Makkot. It reads this way:

> Rabban Gamaliel, Rabbi Eleazar ben Azariah, Rabbi Joshua and Rabbi Akiva were coming up to Jerusalem together, and just as they came to Mount Scopus they saw a fox emerging from the Holy of Holies. They fell a-weeping and Rabbi Akiva seemed merry. Wherefore, said they to him, are you merry? Said he: Wherefore are you weeping? Said they to him: A place of which it once was said "And the common man that draweth nigh shall be put to death" is now become the haunt of foxes, and should we not weep? Said he to them:
>
>> Therefore am I merry; for it is written, "And I will take to Me faithful witnesses to record, Uriah the Priest and Zachariah the Son of Jeberechia." Now what connection has this Uriah the priest with Zechariah? Uriah lived during the times of the first Temple while Zechariah lived during the second Temple; but the Holy Writ linked the prophecy of Zechariah with the prophecy of Uriah. In the prophecy of Uriah it is written, "Therefore shall Zion for your sake be ploughed as a field" [i.e., be left as wasteland]. In Zechariah it is written, "Thus saith the Lord of Hosts, There shall yet sit old men and old women in the broad places of Jerusalem," so long as Uriah's prophecy had not had its fulfillment, I had misgivings lest Zechariah's prophecy might not be fulfilled; now that Uriah's prophecy has been literally fulfilled, it is quite certain that Zechariah's prophecy is to find its literal fulfillment.
>
> Said they to him: Akiva, you have comforted us! Akiva, you have comforted us![7]

Staring at the two old people, wizened by age and, no doubt, the worries that had tormented their generation of Jews (for they were clearly European survivors), yet seeming so at peace with their surroundings on the square, I was struck hard by the literalness of the prophecy. I walked back to the hotel quaking inside and overwhelmed by the fact that I could personally testify to the validity of Rabbi Akiva's optimism two thousand years earlier. The thought of that incident still sets my head spinning, and the faces of those two anonymous old people continue to gaze at me between the lines on the computer screen as I commit that memory to paper.

5

Preparing the Lavi
Report

I was soon thrown back into the far more mundane business of amassing facts and figures. We were not doing very well. While we obtained some material, we still were frustrated by the Israeli refusal to disgorge key factors like the work breakdown structure. I knew we would be back in Israel soon, and the Israelis did not relish that prospect at all.

The first evidence of Israeli stonewalling came within days of our arrival home in the form of a very polite note from Mendy Meron, Rabin's director general at the ministry of defense, to Fred Iklé. Meron, a true gentleman, opened by expressing his appreciation for "the very high level of professionalism and candid and friendly exchange of views and data that took place during the visit." (The description of the exchange of views, by the way, is also diplospeak for a knock-down drag-out battle.)

He went on to describe Israel's "somewhat uncomfortable situation" regarding the second phase of the cost estimate, that is, our follow-on questions. I had pressed for answers by mid-September, so that I could have the report complete before the end of the year. The Israelis desired more time to formulate their answers, and Meron requested a delay until mid-October. He claimed that our "very long and complicated questionnaire" required much preparation, "some of it completely new, because we do our cost estimate in a different manner." He added that the people who would reply to our questions were the same ones directing the aircraft program.

Meron therefore asked that we permit his people "to make those . . . preparations not under pressure. . . . We will fulfil our commitment," he concluded. "We may not be able to do it always within the required timetable but we will do it."[1] I did not doubt Meron's sincerity; in any event, we had little option but to wait until the Israelis supplied us with the information we requested.

Fred, as was his wont, responded most courteously to Meron. He acknowledged the pressure on the MoD staff, but emphasized "the need for prompt action to gain a thorough understanding of the Lavi program and its most likely future costs." He therefore proposed that the next visit take place on October 7 and that it last for ten days. He concluded that "to delay any further would only force both of us to make defense resource allocation decisions without the benefit of the best available information. This may result," he added somewhat archly, "in wasted resources, a situation both of us can ill afford."[2] Meron must have breathed a sigh of relief at the two-month delay, and hurriedly wrote back that the proposed dates were acceptable to him.[3]

Shortly before Fred had penned his reply to Meron, *Aerospace Daily*, a trade newsletter, carried one of the first detailed reports of our review. I had provided the reporter with some background information, and its description of our efforts was reasonably accurate. What I found most interesting was the Israeli reaction to our work. One official—no one was prepared to go "on the record"—stated in response to a question about our critique of the Lavi's costs that "we've heard all that stuff." Another contended that the program's problems were "minor," while a third conceded that "correcting defects . . . means spending additional money," without indicating where that money would come from.[4]

These face-saving statements were to be expected. Program managers rarely, if ever, freely acknowledge the difficulties they are encountering. Far more insidious were a different set of reactions that I encountered within forty-eight hours of my arrival home. Phil Merrill, who had been on the April trip, told me in the Pentagon that he had heard from his good friend the recently retired Ambassador Sam Lewis that the Israelis were complaining about the Pentagon's "patronizing" attitude. How could DoD send an Orthodox rabbi, not a real defense expert, to tell them what to do?

I was extremely upset. I had told myself over and over that I would be the subject of personal attacks. But when they began to come, they hurt no less for having been anticipated. The day after Phil passed along his story, I was visited by a Defense Department employee who had very close ties to the Israelis. He began to ply me with questions regarding my personal motives for pursuing the Lavi case so vigorously. I had known him for some time; he was a former academic who had joined the department as a Middle East specialist. We were on good enough terms for me to tolerate his sitting in my office for an hour, pumping and pumping me. Did I have a "hidden agenda"? Did I bear some grudge against Israelis? Was the F-16 really a better airplane? Did I realize that I could cause Israel much damage if it became clear that the government had manipulated the Lavi's costs? The questions kept on coming as relentlessly as the waves of a summer high tide. I tried very hard to keep my cool; inwardly I was more than a little agitated.

I hardly felt better when another friend with excellent contacts in the Israeli navy told me that someone who knew about our friendship asked him "whose side" I was on. What was going on here? Yesterday I was the incompetent rabbi, who had no business mucking around in the province of economists. Today I was the self-hating Jew, with some ancient ax to grind against the Chosen People's chosen state. It hurt, and hurt badly. I determined that, come what may, I would never let on just how much it hurt.

All through the early fall my team pressed forward with its analysis of the costs and figures that the Israelis provided to us. I was preoccupied with other matters, the *Defense Guidance,* the Anglo-French competition to sell the U.S. Army a costly mobile communications system, and, most urgently, a project to prevent the closure of the naval station on the small Caribbean island of Antigua. There were also the High Holidays, not only New Year's and Yom Kippur, but also the Sukkot festival. Sukkot—or Tabernacles—is, like Passover, an eight-day holiday with four holier days that allow for no workaday activities. Each year I would have to miss at least seven working days. Since the holidays were spaced over a four-week period, when weekends were taken into account, I rarely was able to accomplish as much as I would have liked to in the month of September, even though I spent many more hours at work on the days that I did make it into the office.

Despite my diversions, we had undertaken one major initiative in early September. We had begun to plan the next phase of our strategy, namely, to develop and analyze alternatives to the Lavi. None of us was under any illusions that such an exercise would be anything less than explosive, certainly in Israel and possibly in Washington as well. Yet, if we were going to assert, as it became increasingly clear that we would, that the Lavi was "too expensive," we had to indicate what was not "too expensive" for Israel to acquire.

We recognized that in order to put together a credible report on alternatives to the Lavi, we needed the cooperation of the major American defense contractors who produced aircraft that might compete with the Lavi for the Israeli air force's modernization funds. To that end, we decided upon a series of team visits to General Dynamics, McDonnell Douglas, Grumman, and Northrop. These visits were termed, in classic Pentagonese, "pre-alternative contractor visits." At this stage, all we wished to do was to assess which contractors were seriously interested in working with us to develop alternatives to the Lavi. We conducted the alternatives effort on a low-key basis; it was the wrong time to draw anyone's attention to it.

As we continued our work on the Lavi cost issue, the department was engaged in a series of intermittent skirmishes regarding the way in which American money was being expended by the Israelis in connection with the program. One such clash took place in mid-September, over Israel's use of foreign military sales credit funds to finance a value-added tax levied on Lavi in-country production by the ministry of finance. From our perspective, this was another case of Israeli chutzpah. It was not enough for the Israelis to have received American taxpayers' dollars for the purchase of Israeli products by the Israeli government. Now the government was levying a tax on those purchases!

Phil Gast, DSAA's director, protested Israeli behavior in a letter to Abraham Ben-Joseph, the veteran chief of Israel's purchasing mission in New York, the largest of its kind anywhere. Ben-Joseph was no innocent, having also served as director general of the ministry of defense, and Gast pulled no punches as he asserted that the effect of the tax was to "take FMS credits provided for the purchase of defense articles and services and permit your Finance Ministry to utilize them for other purposes." Gast then informed Ben-Joseph that the funds were meant to be only for defense purposes, and their diversion to VAT would not be authorized after October 1.[5]

I often wondered whether the Israelis ever really understood just how much they infuriated Americans by their behavior, and how much damage such petty nuisances did to American goodwill. Not only did they reinforce the views of those who harbored prejudices against the Jewish state; they also confirmed the often queasy suspicions of other, more objective public servants that somehow they, and the American public, were being ripped off by Israel. Was the money that VAT brought in worth the suspicions it bred and reinforced? I doubted it. Perhaps the Israelis thought otherwise.

It was also in September that Fred Iklé decided to promote me to the rank of deputy under secretary of defense, a notch higher than my assistant under secretary status. I had found that the shade of difference in rank had a significant impact on my dealings with the military during the arduous give-and-take of budget negotiations. Uniformed officers are always sensitive to the nuances of rank. Ideas and arguments, however cogent they might be, often go nowhere unless underpinned by rank. The Israelis, oblivious to anything but their own interactions with me, immediately assumed that I had been promoted as a reward for giving them a hard time over the Lavi. It did not help that the promotion came through only a few weeks before I was again to visit Israel in November.

Early in October, Foreign Minister Shamir came to Washington on one of his frequent pilgrimages to the States. I chatted briefly with him at the reception that the embassy threw in his honor. It was a crowded affair, filled with government officials and prominent American Jews, many of whom had flown in from New York especially for the occasion. Shamir asked for my father and mentioned that he hadn't seen him or my uncle in many years; he had a twinkle in his eye as we conversed in Yiddish. He didn't mention the Lavi at all, nor did I.

By the time Shamir visited Washington my team had made enough progress for us to begin holding a series of meetings to determine our strategy for dealing with the Israelis. The cost estimates were beginning to take shape, and they were unlikely to give the Israelis any comfort. We were going back to Israel again, to brief the MoD and IAI team on our progress. Our estimates had come down, as I had expected them to, once the Israelis began providing us with information. But just as the Israelis continued to withhold data—notably the critical work breakdown structure—so were our estimates still much higher than theirs.

We did not expect to glean much more on our visit. The department's attitude was best summed up in a one-word change that Fred

Iklé scribbled over an Israeli headline that was reported to us in a cable from Embassy Tel Aviv a few days before our departure. The headline (from the newspaper *Davar*) read, "United States May Ask Israel to Reveal Aircraft's Secrets." Fred had crossed out the word "secrets" and scribbled "costs!" in its stead.

We departed for Israel on Tuesday, October 15, 1985. My plan was to spend a few days there and then attend some meetings on NATO research and development cooperation in Paris on the following Sunday, October 20. The rest of the team would stay on in Israel until the end of that week to continue their fact-finding and discussions.

We met with the same cast of characters as during my initial trip with Iklé in April: Eini and his IAI/MoD Lavi team and economic adviser Zvi Tropp (with whom I had already met twice before, in Washington in June and in Tel Aviv in July) at the defense ministry, Modai at the finance ministry, Neubach in the prime minister's office. We also met with David Ivri, the chairman of Israel Aircraft Industries, who would later replace Mendy Meron as MoD director general. Ivri was a former commander of the Israeli air force and, like Rabin, a genuine war hero. Short and taciturn, with penetrating ice-blue eyes, he usually left me feeling very uncomfortable after we met. Needless to say, given his position at IAI, I had no doubt where he stood on the Lavi issue; he was a formidable opponent, and remained so well after he left IAI for the defense ministry.

Finally, I met with Defense Minister Rabin as well. I had begun to enjoy my meetings with him (we had met again briefly during my July visit) and had made friends among his staff as well. For his part, he seemed increasingly relaxed with me when we met. We always conversed in English, not only out of consideration for Tom Picker-ing, who usually accompanied me to these meetings, but also, as I discovered on this particular visit, because Rabin preferred it that way. I could have managed in Hebrew, and certainly would have had no problem understanding his Hebrew. Nevertheless, whenever I broke into a brief Hebrew phrase he would ignore it, and continue in English as if the conversation had never shifted linguistic gears.

My meeting with Rabin took on a special significance that in my view went well beyond the case of the Lavi. To be sure, I gave Rabin an update on where we stood regarding our cost analysis, and elicited little by way of reaction. He was, as they say in the Penta-gon, in a "wait-and-see-mode." But I also surfaced the question of Israeli naval modernization. I suggested that a coordinated American-

Israeli effort to evaluate the costs of Israel's proposed submarine and surface ship acquisitions might enable both countries to avoid the mess that the Lavi program had created.

I was hoping Rabin would agree. For me, his acquiescence would be the clearest signal yet of his intention to approach the Lavi issue objectively, which meant that there was a chance he would terminate the program. Still, I wasn't sure. Rabin was a secretive man, and I wondered whether he would be worried that word would leak out of a new naval modernization study that would lead observers to believe he was working with us hand-in-glove on the Lavi as well. But Rabin came through. He stated that he thought a cost estimation study was a useful idea, though he would not commit himself to its findings, which was natural enough.

I knew the Israeli navy would be delighted. Its commander, Avraham Ben Shoshan, was a capable, outspoken fellow who recognized that it was in his interest to enmesh the Office of the Secretary of Defense in his work, if only to offset the suspicions of both the United States Navy and the Joint Chiefs of Staff, which reflected the navy's view.

Israel's naval situation was actually somewhat ironic when compared to the Lavi. The Israeli navy, long the poor cousin of the Israel Defense Forces, consumes only about 5 percent of the defense budget. In 1982, after considerable lobbying by the Israeli navy, the ministry of defense had formally proposed the initiation of a major naval modernization program, incorporating both the design and construction of a new class of corvette—a surface ship that was larger than any of the gunboats in the Israeli navy—and a new class of diesel-electric attack submarine. The Israelis were operating three rather ancient and quite small British-built Gal class submarines. The navy was especially sensitive to the need to maintain modern subs, and not only for reasons of military capability. In 1968, Israel had lost a submarine, the *Dakar*, somewhere off Egypt. Its disappearance had never been fully explained or understood, and the country had mourned the lost sailors. In fact, the loss had raised a religious issue as well. Under Jewish law, a woman whose husband has disappeared but is not proved to have died cannot remarry. She is termed an *agunah*, "chained one." The *Dakar* left many such *agunot*; the Chief Rabbinate ruled, however, that the men should be treated as dead, thereby permitting the women to remarry and try to put their lives back together again.

While the proposal for an upgrade of Israel's surface fleet did not engender much controversy, it was not at all clear whether Israel could afford to fund the construction from its own resources, and American dollars for Israeli purchases in Israel were being directed entirely toward the Lavi program. The Israeli navy began to look to the United States as potential builder of at least some of its boats, presumably with military assistance funds.

In contrast to the surface ship modernization program, the plan to replace the Gal subs involved some very sensitive, indeed controversial, issues that went far beyond the matter of whether funds were available for its implementation. The Israeli navy preferred to buy American; but the U.S. Navy, even with the pro-Israel John Lehman firmly in control as secretary of the navy, refused to tolerate the construction of diesel subs in the United States, even if only for export. Diesel submarine construction posed a threat to the navy's nuclear barons—the submarine fleet at the time was over 90 percent nuclear-powered. They feared that if diesel subs, which cost about 25 percent of a nuclear-powered submarine, were built in the United States, congressional critics of the nuclear-powered program would soon legislate that the navy also acquire diesel boats.

The Pentagon was well aware of the Israeli desire for an American diesel boat; its response was to question the validity of the Israeli requirement. A Joint Staff study visit to Israel concluded with a report on February 13, 1985, that the Jewish state did not need to replace its Gal force. The Americans rejected the Israeli claim of potential structural failures in the Gal boats. Moreover, they questioned Israel's need for a larger submarine and the Israeli requirement for greater endurance. Privately, U.S. Navy officials speculated that the Israelis were seeking a submarine to launch special operations forces at great distances, possibly against Libya. In all events, the report concluded that the Israeli navy could wait at least another decade before acquiring new submarines.

The Joint Staff report had been a crushing blow to Israeli hopes. There appeared few other options for the Israeli navy, other than the possibility that the submarine could be built in either Germany or the Netherlands—if the funds could be found to pay for it. In April 1985, however, shortly before our first Lavi trip, John Lehman had been to Israel for a brief visit of his own and had hinted broadly at a joint press conference with Rabin that a solution had been found. Israel would build its submarines in the Haifa yard and its surface ships in the United States. Lehman gave no specifics, however, nor could he, for the details of the program, and its costs, and indeed Cap Weinberger's authorization for such a plan had yet to be obtained.[6] Moreover, the navy secretary's commitment did not necessarily imply the endorsement of the uniformed navy. The "blue suiters" sympathized with the Joint Staff's views and considered that Israel did not really need to replace its Gal boats at that time.

The Israeli navy was not certain that the Haifa yard actually could build the submarines, and in any event was under no illusion that Haifa could design them. Accordingly, it hedged its bets by engaging the German engineering firm of Ingenieurkontor Lubeck (IKL) in July to design the boat. Buying a submarine from Germany, or indeed building it in Israel, raised the same difficulties that building a new surface ship in Israel did: there wasn't enough money in the Israeli defense budget. The German option posed even more difficulties, however. There was always an outside chance that the Congress could be persuaded to allow Israel to spend more offshore funds for its programs, and thereby fund the submarines (or surface ships) with American security assistance dollars. There was less likelihood that Congress, much less the administration, would go along with a proposal to spend American dollars for Israel *in Germany*. That bridge could be crossed in the future, however, and the decision to have the Germans design the boat did not necessarily prejudice a subsequent decision as to where the subs should be built.

John and I had spoken about the Israeli program upon his return, and I was certain that Rabin had been made aware of the fact that I shared Lehman's views about the need for new submarines, though I was publicly agnostic about where they should be built. Privately, I had felt for many years that the U.S. Navy was short-changing the U.S. shipbuilding industry by not allowing diesels to be built in the United States. All we were doing was giving away a lucrative market to the Europeans. We were also depriving ourselves of new technologies that enabled nonnuclear submarines to travel quietly at higher speeds while submerged for longer periods than ever before. I did not, however, raise these issues in the context of our subsequent naval analysis.

In any event, Rabin's giving me the green light to evaluate another aspect of his defense program added momentum to the plan that he and Lehman had agreed upon and paved the way for a concerted attempt to win over OpNav, that is, the U.S. Navy staff, to the Israeli program. In addition, by agreeing to the study, Rabin had also legitimated our ability to analyze Israeli programs, even if he was not prepared to accept our results on the Lavi. I returned to the United States feeling that we had turned a corner even before our final analysis of the Lavi had been presented to the MoD. About eight weeks later, and after considerable coordination with John Lehman—who had once been Fred Iklé's deputy at the Arms Control and Disarmament Agency—Fred formally assigned me the responsibility for producing "an analysis of the cost and budgetary implications of the Israeli submarine request and its concomitant proposal regarding missile boat construction in the U.S."[7]

My final series of meetings in Israel had taken place on Sunday morning, just prior to my departure for Paris. The trip had therefore once again afforded me the opportunity of spending Shabbat in Jerusalem, and again I stayed at the Laromme. On this occasion I walked up Keren Hayesod Street to attend services at the Jerusalem Great Synagogue, which is colocated with the building that houses the offices of the Israeli Chief Rabbinate. The synagogue was built by the British magnate Sir Isaac (later Lord) Wolfson, in the style of the ornate London synagogues. Termed "Goren's cathedral," after the late chief rabbi who frequently preached there, it is a mighty monument to the worship of Heaven. One ascends a wide staircase, passing a small chapel that houses a Sephardi synagogue, which often plays host to Rabbi Ovadia Yosef, one of Israel's leading authorities on the Halacha (Jewish law and practice) and at one time Rabbi Goren's counterpart as chief rabbi/*rishon l'Tzion*—literally "first in Zion"—of the Sephardim.

At the top of the staircase one enters a huge sanctuary. It is dominated by an impressive chandelier and filled with polished wooden pews that face an ornate Ark holding a multitude of Torah scrolls adorned with silver crowns and breastplates. Behind the Ark, on the eastern wall, is a Chagall stained-glass window, depicting in bold colors the glory of the Lord. The elevated platform where the Torah scrolls are read stands in the center of the synagogue, in accordance with Orthodox tradition. Before the Ark is another raised platform where the choir stands.

The synagogue is a favorite of visiting Americans of modern Orthodox bent. It was something more for me, however, reminding me of many happy Saturdays attending synagogue in London, including Sir Isaac's own—and very grand—Central Synagogue. Indeed, I had heard the Great Synagogue's cantor, when he still performed (for he didn't just lead the service, he performed it), before the congregation of the very imposing Finchley Synagogue in North London some fifteen years earlier.

I must admit to being a very high church Jew. I enjoy the grandeur of a large synagogue, including the balcony for women, modeled, so we are told, after the Ezras Nashim, the women's chamber in the Temple. I take pleasure in hearing the harmonies of cantor and choir, intoning melodies of a service whose text has not changed for many generations. On the other hand, I feel cramped in the smaller shuls that often dot Jewish neighborhoods in America's older cities. Far from considering it an insult to Rabbi Goren, I have always felt that his being linked to the "cathedral" was a mark of his good taste. For as the Scripture has put it, "in the multitude of the people is the king's glory."[8]

Much as I wanted to turn my attention to the Israeli navy, not least because I had first entered the defense business as a naval forces analyst for the Congressional Budget Office and still very much loved to deal with issues affecting the maritime services, I recognized that I still had to address first things first. I couldn't focus on the Israeli navy until I was done with the Lavi.

Within days of our arrival home from Israel, I found myself dealing with a new twist to the Lavi story. The press was reporting the possibility that the Lavi might replace the A-10 ground attack plane in the U.S. Air Force inventory. Both the weekly *Defense News* and the October 18, 1985, *Inside the Pentagon*, the maiden issue of yet another daily newsletter, had announced that, as the latter put it, "top Pentagon policy and r&d [research and development] officials are seriously considering the possibility of a major U.S. buy of the Israeli Lavi as the next USAF close air support fighter."[9] It was true that a replacement to the A-10 was beginning to be considered within the air force and the Office of the Secretary of Defense. That, however, was where the true part of the story ended. The bit about the Lavi was a pure fabrication, no doubt circulated by the plane's sympathizers.

The A-10, nicknamed the Warthog because of its relative ugliness, was most certainly not an air force favorite. It had been built around a very powerful 30-millimeter gun and was meant to be a tank killer.

The flyboys, enamored of high-powered jets such as the F-15, as well as with air-to-air combat, at which the F-15 excelled, had never hidden their disdain for the relatively inexpensive, slow-moving Warthog. Not suprisingly, there was an undercurrent of support for dropping the plane from the air force inventory, and, no doubt, those who fanned the rumors of its replacement by the Lavi were counting on that dissatisfaction to manifest itself in support for the Israeli plane. Still, however much the air force may have disliked the A-10, it liked the Lavi even less, as I quickly learned from several phone calls I made to a variety of officers on the day the second story appeared. Nevertheless, once the rumor had circulated, it was impossible to put it to rest. As one air force official told *Defense News*, anything was possible.[10] That slim will-o'-the-wisp was enough for the Lavi supporters, not least among them Moshe Arens, to delude themselves that the United States Air Force ultimately would be the *deus ex machina* for the Lavi.

The A-10 was not dropped from the air force inventory. In fact, it proved itself far superior to what its detractors had expected. It performed remarkably well during Operation Desert Storm, and was responsible for a very high proportion of the destruction wreaked on Iraqi tank forces. That was all far in the future, however. As a result of the Lavi-for-Warthog rumors, I found myself wasting lots of time and energy over the next eighteen months denying over and over again to the press that the United States had any intention of buying the Israeli plane.

As we approached the report's completion, we were heartened by a significant provision in the version of the foreign assistance appropriations bill that was soon to be addressed in the Conference Committee of the two houses. As DSAA's Glenn Rudd had alerted us in April, for the first time, the language dealing with foreign military credit sales stated that "not less than $300,000,000 shall be for the procurement in Israel of defense articles and services, including research and development, for the Lavi program *and other activities* [my emphasis] if requested by Israel." This was a breakthrough. Previously, all offshore procurement had been tied to the Lavi. Now the funds were open to competition from other critical Israeli programs. In fact, the Israeli press was reporting that the economic section of the Israeli embassy in Washington, headed by Danny Halperin, one of the town's shrewdest operators, had pushed for the congressional language in case the Lavi never flew.[11]

Regardless of who was actually behind the new congressional action, it meant that advocates of land and naval forces modernization, and even advocates of other programs within the Israeli air force, would view the Lavi as swallowing up not only Israeli funds that they coveted but also American funds that they previously had thought were unavailable to them. It also clearly meant that the Congress was not as strongly supportive of the Lavi as Arens and his henchmen asserted. Perhaps the Congress was not prepared to oppose the Lavi, but its support seemed increasingly lukewarm.

It was clear to me at least that the debate in Israel would soon take an even sharper turn. I expected that more of my hitherto silent allies would be coming out of the closet as the implications of this provision—which I was certain would pass the Conference Committee—sank in. Thus, as the report neared completion, the question of managing its public affairs aspects—i.e., publicity—became an increasingly urgent subject in our strategy sessions.

I had continued to keep in touch with Jewish leaders such as Tom Dine of AIPAC, letting them know, at least in general terms, of the progress we had made. With the concurrence of Fred Iklé and of the other powers-that-be in the Pentagon, I also had begun to give backgrounders to the press earlier in the summer, both to the trade press and to certain reporters in the Jewish press, notably Wolf Blitzer, then the *Jerusalem Post*'s Washington correspondent. I had dealt with Wolf before and knew I could trust him. These backgrounders attracted relatively little press coverage, which I felt was just as well, since premature publicity could damage our effort. On the other hand, by keeping the press informed, I had ensured that it would at least give me a sympathetic hearing once the time came to "go public." Now was that time; we needed to spread the word that there was something fundamentally wrong with the Lavi program.

The first American report to discuss our work in detail was a lead article in the November 18 issue of *Defense Week*, a journal that made the rounds of the defense community. Opening with the assertion that "Israel's Lavi fighter program could be heading for huge cost overruns," the piece went on to recount the cost estimates that both we and the Israelis had put forward before the summer. These estimates, which were reported as approximately doubling the Israeli projection of $9 billion, were higher than the numbers our team were preparing for final presentation in Israel; our own estimate was closer to $14 billion. Nevertheless, it was the fact of the discrepancy between the two sets of estimates, rather than their actual details, that was of considerable significance.

The story did not end with our cost analysis; in fact, that was only a part of the Lavi's growing troubles. *Defense Week* reported that despite increasing costs, the Israelis were having difficulty absorbing American funds that had been dedicated to the Lavi program, leading some observers to conclude that the Lavi was simply a cover for Israeli use of American dollars for a variety of efforts. The article also noted that former defense minister Ezer Weitzmann (now, like Moshe Arens, a minister without portfolio) had come out against the plane in a cabinet vote held in August. We soon learned that Finance Minister Modai had joined Weitzmann in voting against the Lavi.

Despite all the negative publicity that the article gave to the plane, the Israelis drew considerable solace from one key observation: it reported that State and Defense were both resigned to the project's going ahead, despite opposition in both departments. Someone at State was quoted as saying, "The battle has been fought and we lost."[12]

In fact, Weinberger and Iklé were not at all resigned to what the Israelis continually presented as the Lavi's inevitable emergence on the scene. Resignation was, however, the prevailing mood at State. Indeed, I was led to believe by several friends at State that Secretary George Shultz really did not want to go toe to toe with the Israelis on this issue. He was working closely with Shimon Peres to rescue Israel from the threat of hyperinflation and did not want to complicate this effort with a quarrel over defense economics that Weinberger's department had initiated.

Shultz's attitude posed another potential complication for me. While his staff might sympathize with our opposition to the program, and indeed was very much involved in our effort, we somehow needed to have Shultz aboard, if only to maintain a common front vis-à-vis the Israelis. Thus far he had not voiced any objection to what we were doing. I could only hope that his staff, especially the people in the Bureau of Politico-Military Affairs (known in State as P/M) who were most involved in our effort, as well as those of the Near East Bureau, would continue to ensure that Shultz not only did not try to quash it, but also would not even intimate to anyone that he had reservations about it.

In mid-November 1985 it was Defense Minister Rabin's turn to visit the United States. I chatted with him briefly after he delivered a public speech on November 15. There really wasn't much for me to tell Rabin at that stage. We were still awaiting final data from the Israelis. Moreover, once completed, the report required Weinberger's approval, and that of the other government agencies involved. I certainly was not going to complicate matters by short-circuiting normal government processes in a matter as sensitive as this one was.

On the other hand, it was important that Weinberger convey to Rabin that he stood foursquare behind my efforts. His military assistant, Major General Colin Powell, asked that I prepare "talking points" for his private meeting with the defense minister. I included in the "talker" Weinberger's hope that we would receive the remainder of the material we had requested within the next three weeks, as had been promised to us. I also added, in parentheses, some words about our looking at alternatives, if the discussion with Rabin moved along well.

In the event, it appears that Weinberger never did surface the question of alternatives, though he did discuss the Lavi in private with Rabin. Nor did we ever receive all the data we requested from the Israelis. I ordered a series of point papers from the staff that indicated just what the Israelis had provided to us and what they had not, and what the implications of the lacunae were likely to be for the analysis. The staff concluded that the Israelis had provided them with enough data on the plane's weapons system to submit it to a thorough analysis. On the other hand, critical performance data relating to thrust, range/payload, and other key parameters had not been furnished to us. The performance characteristics were required to determine the overall probability of design changes and indicate risk areas, both of which would affect cost estimates. Our analysts were forced to fall back on what they termed "a limited, subjective," and therefore unsatisfying performance evaluation.

Perhaps it was not surprising that the Israelis did not want us to assess the probability of design changes. These were taking place even as we were wrapping up our assessment. For example, on November 27, Phil Gast received a request for the release of new funds to support a major wing redesign of the Lavi. Grumman, the contractor producing the wing and vertical stabilizers, had already been awarded contracts totaling $127.3 million. Ben-Joseph at the New York purchasing mission was now asking Gast for an additional $46.5 million—a whopping 38 percent increase, and even that sum may not have fully accounted for the cost of the change.[13]

Grumman's submission to IAI for the increase was instructive.

In forwarding its proposal to IAI on July 3, Grumman had stated that "based on improvements in weight and performance that we have seen to date, the decision to redesign was well founded. At the time the decision was taken both parties did not realize the magnitude of the change in either engineering or tooling."[14] No wonder the Israelis were not helpful when it came to projecting possible cost increases that were due to changes in the contract; they were in the midst of incorporating such an increase at the very time we were requesting our data!

We never did obtain a number of other key cost parameters either. These included a detailed engineering weight statement; a statement of material mix that supported the weight statement; full-scale development man-hours; priced bills of vendors' materials; "wraparound" or "wrap" rates (total hourly rates that cover direct, indirect, and administrative expenses); the master scheduling plan; much of the IAI's cost history relating to the Kfir airplane; and details of the integrated logistics system.

Our inability to obtain these parameters had a major impact on our cost estimates. The weight-related figures were required for determining the airframe unit weight, a basic element in cost models. Schedules were required to enable our team both to assess the time required to perform tasks and to determine that tasks performed were on schedule. They also were needed as input to our manpower projections and their costs. The wrap rates were necessary for converting forecast work hours into dollars. The support plan for the logistics system was required to establish the cost of operating the Lavi once it was in the Israeli air force's inventory and to estimate the cost of spares. The IAI history was required for us to adjust the U.S. manufacturing capability inherent in our cost database to an Israeli capability.

We all doubted that the Israelis really did not have these factors. We wondered how they could produce a plane without them. And we certainly were not surprised that when the Israelis later attacked our report, they used arguments—such as the "cheaper" methods Israelis employed for building aircraft—that obviously would have been unavailable to them had they provided us with the information we sought.

Throughout December 1985, the team crunched the numbers it did have, and by the end of the month the Lavi report was virtually complete. I wrote the executive summary on my own, having long ago been convinced that a five-page executive summary is at least as important as the five-hundred-page study it might precede, and certainly would have a wider readership among more senior people. For that reason too, I convinced Iklé that the summary should be unclassified, so that it could reach the largest possible audience.

Ironically, I actually would have had a better chance to garner more headlines had I classified the report. Journalists, at least those in the national security arena, tend to pay less attention to documents that are not marked SECRET, at the very least (TOP SECRET is preferable), whatever their contents might be. Still, I did not want to complicate an already controversial issue with a major witch-hunt over leaks, and I knew that even if unclassified, our report would tell quite a story.

In the meantime, the controversy surrounding the plane was increasing daily. Modai's finance ministry had proposed both an additional cut in the defense budget and a plan to seize upon the congressional language, now fully approved, to reallocate $250 million of the available $300 million in offshore funds to programs other than the Lavi. Modai himself was pronouncing to the press that the plane would "fall by itself" and that "in consideration of the U.S. Administration's approval for Israel to convert $250 million of American aid into shekels to be used on projects other than the Lavi, there is no longer any reason to spend the money on the fighter, which Israel does not possess the financial means to support."[15]

Senior think-tankers in Israel, no doubt sensitive to the new import of the congressional action on offshore procurement, were openly pronouncing that the Lavi was unaffordable. "Even if we build the plane," said one, "it's hard to see how the Israeli air force could afford to buy it without mortgaging other weapons development projects."[16] The Israeli military was equally quick to appreciate the implications of the Congress's action. No less than Amos Lapidot, air force commander and a staunch Lavi advocate, was quoted as saying that the planned Lavi production of thirty aircraft annually would "cut severely into . . . the Air Force budget, stunting other developments in the force."[17]

Yitzchak Rabin remained somewhat more vague about his views. According to the well-informed independent daily *Ha'aretz*, Rabin told a treasury meeting early in December that he would rather kill or significantly cut back the Lavi program than sustain additional budget cuts.[18] His argument was part of an impassioned plea for avoiding additional defense cuts, which made it seem as if he still supported the project. For that reason, the significance of his remark was not fully comprehended by the Israeli press or public. In fact, however, there was little doubt that the defense budget was due for additional reductions, in the next few years even if not in 1986. Rabin had in effect served notice that the Lavi was more vulnerable than ever.

Early in January we began our round of internal briefings, providing various DoD officials with what we called "the numbers." The finishing touches were put on the final draft of the report on January 22. We knew that the Israelis were bracing themselves for our visit, and they had every reason to be apprehensive. The pressure on the defense budget and Rabin's pronouncements meant that the Lavi's backers were already fighting for their lives. The last thing they needed was our exercise.

The report was entitled *The Lavi Program: An Assessment of Its Mission, Technical Content and Cost.* Its main body was but twenty-three pages, but it also included nine appendices, one of which, "Technical Data Summary and Assessments," ran to 140 pages alone. Other appendices, many of them unclassified, provided a virtual history of the entire process of analyzing the program: one outlined our cost-estimating methodology, another our cost-estimating process, a third our data request. If not unique, the study was certainly a rarity: a detailed evaluation of every aspect of a foreign state's weapons program that, unlike other classified analyses, was open to that state for inspection.

The report gave the Israelis their due. It validated the mission requirement; in other words, we did not dispute the Israelis' claim that they needed a modern combat aircraft to replace their A-4s and their older F-4s. We also noted that the main risks related to potential schedule delays and cost, and not to performance. While we had some concerns about some of the performance specifications, for the purposes of the report we accepted the Israelis' description of what the Lavi could do. That meant that we accepted the Israelis' claim that the highly maneuverable plane could travel nearly twice the speed of sound (Mach 1.8). We acknowledged both its air-to-air and air-to-ground capability, though we had far more material to support the plane's ability to carry out the former mission than the latter. We did not dispute IAI's claim that the plane had a significant combat radius for a fighter (the longest distance it could fly and still return without refueling): about 1,000 nautical miles (nm) in the air-to-air role, and 600 nm when flying a demanding low-level air-to-ground mission.[19]

Still, the report pulled no punches. Its very first page asserted that "Israel has seriously underestimated the cost of the LAVI program. Certain cost elements, such as those for the production tooling, have simply not been accounted for; while others, such as labor rates and fuselage cost, are based on uncertain methodologies."[20]

The cost differential was large. Our estimate of the plane's "flyaway cost," that is, the cost of each plane off the production line, was nearly $7 million greater than the Israeli estimate of $15.2 million, a difference of over 45 percent. If the initial spare parts and special ground support equipment were added to the flyaway cost—creating the more meaningful though less frequently quoted "production cost" category—our estimate was over 52 percent higher.

We emphasized that our unit cost estimates depended upon the Israeli production profile, which we had accepted without question. This profile called for maximum rates of two aircraft a month and a total program of three hundred aircraft. Were the program to be "stretched"—that is, spread over a longer time frame due to budget constraints, as was usually the case with American aircraft programs and most of our allies' programs as well—the unit cost of the plane would inevitably rise.[21]

Perhaps our most startling revelation, however, was the additional cost burden that the Lavi was likely to impose on an already stretched Israeli defense budget, and on the U.S. military assistance program that supported it. We predicted that by 1988 Israel would require an additional $766 million to support the Lavi. By 1990 that figure would rise to $937 million, representing 49.7 percent of the entire U.S. assistance budget. There was simply no way that the United States could underwrite these additional costs, which altogether were predicted to total $16.4 billion over the life of the program.

We argued that the American budget environment, which had already led to a $77 million reduction in fiscal year 1986 assistance to Israel as a result of the Gramm-Rudman-Hollings budget legislation, simply would not tolerate future increases in that assistance. As a result, we stated, "other high priority Israeli programs toward which the U.S. budget contributes might have to be reduced or cancelled."[22]

In pointing out the risk to other Israeli programs, and doing so in the fully unclassified executive summary, we were deliberately reaching out for new support within the Israeli military-industrial complex. My colleagues and I had begun to hear stories about Israeli engineers being laid off as a result of funds flowing solely to the Lavi program. Major defense firms other than IAI were hurting. The Merkava tank program people, the Israeli navy's complex of contractors, and the military officers associated with both these programs bitterly resented the prospective domination of future Israeli budgets by the Lavi and its creators at IAI.

As if our current estimates were not enough to cause consternation throughout the Israeli defense establishment, we further argued that our numbers were probably conservative. For example, we assumed "learning curves"—the savings generated from repeating identical manufacturing processes—to be constant. In fact, the Israelis had postulated that some subsystems would be transferred to Israel after initial production elsewhere. The result, of course, would be a shift in the learning curve as the Israelis had to "learn" the manufacturing process that they had begun to manage on their own. Past experience in the United States and elsewhere, and most notably the Bet Shemesh fiasco, belied the assumption that costs would not rise.

We also had not addressed the cost impact of delays (termed "schedule risk"), which we knew could well be extensive. We had only included a standard factor for engineering-change proposals, which inevitably take place in the development of major weapons systems. Yet for every year that the production schedule would actually slip, the program cost would rise by $250 million, and virtually everyone expected the schedule to slip, since it already had done so. "In short," we concluded, "the cost of the LAVI could well be higher than postulated in this report."[23]

As we continued our round of Washington briefings, word of the report naturally began leaking out in the trade press. Nevertheless, the report's details miraculously remained secret. As for the Israelis, they missed no opportunity to pillory me for every misfortune that befell the Lavi. I learned that Abraham Ben-Joseph, of the New York mission, was blaming me for "holding up" the Lavi funding process. He even went so far as to compare me unfavorably with Phil Gast, with whom Ben-Joseph interacted at least on a weekly basis, and who had long served as the purchasing mission's primary *bête noire*. By now I had accepted my role as Israel's evil demon. I was convinced that the Lavi was nothing else than a massive rip-off that would cost the American taxpayer precious dollars and the Israeli citizen even more precious security. Calumny, I found, came with the territory. What worried me more, apart from the impact that the abuse was having on my family, was whether Israel would make the same mistakes with its naval modernization program, which I considered to be vital to its security and for which—in the case of the proposed new Israeli diesel submarine—there was no American alternative.

6

Israel Responds to the Lavi Report

Our trip to Israel was finally set for the week of February 9, 1986; we were to spend only three days there. On February 4, Reuters reported that Israel might cancel the Lavi for financial reasons; the report did not mention our study, but it did little to earn us a warm welcome in Tel Aviv.[1] The following day Fred Iklé and I once again briefed Senator Rudy Boschwitz on our findings. This time Boschwitz was all ears, and was, in general, sympathetic to our case. By this time too, word of the report was all over the press. *Aviation Week*, perhaps the most widely read defense industry trade journal, carried a major story on the study, including a statement by IAI chairman David Ivri that the Lavi's "flyaway costs" would total between $13.5 and $15 million, which were figures even lower than those the Israelis had officially provided to us.

The weekly also carried a report that Rabin's defense ministry had requested the cabinet either to slow down or cancel the Lavi. Citing an Israel Defense Forces work plan that called for a "reevaluation" of the project, the report noted that Israel had to reduce its budget by $100 million and had few places to cut other than the Lavi. It noted that both Minister Without Portfolio (and former defense minister) Ezer Weitzmann and Deputy Chief of Staff Dan Shomron felt that Israel should consider canceling the program and buying U.S.-built F-16 fighters instead.[2]

Weitzmann's public opposition to the program was well known.

Never the shrinking violet, he asked to see me when I arrived in Israel that week and proceeded to describe the plane's supporters as "full of shit." But Weitzmann was considered a maverick, and it was not at all clear how much his opposition counted for in the inner workings of Israel's policymaking hierarchy.

Shomron's willingness to leak his opposition to the project (or have it leaked—it made little difference whether he did so himself) was quite another matter. As deputy chief of staff, he commanded considerable attention and was likely to exert no small influence over any wavering members of the inner cabinet that dealt with security matters. I had known for some time that Shomron opposed the Lavi. So did Moshe Arens, who later was to bitterly oppose Shomron's promotion to chief of staff. Normally, the deputy chief of staff is virtually assured promotion to the top spot. But I heard that Shomron was subjected to a vicious smear campaign before he received his appointment and that somehow Arens was involved, though he had left no fingerprints behind. If Arens was indeed involved, I was not surprised; he was known to be a tough in-fighter.

Shomron's promotion was far in the future, however, when I arrived in Israel. On the other hand, I was under sustained personal attack. I was constantly receiving secondhand reports of the most recent disparaging remarks about me made by one or another staffer at Israel's Washington embassy. I landed in Tel Aviv feeling rather gloomy.

After a day of trying to catch up on sleep, with only middling success, I began my first real working day at breakfast with program manager Menachem Eini, who was surprisingly polite, and the MoD's Zvi Tropp. We followed by briefing Tom Pickering at our embassy; he asked some very penetrating questions and counseled that we adopt an even, nonconfrontational tone.

We then briefed Mendy Meron, MoD director general, on our findings, which he found nothing less than devastating. I had followed Tom's advice, however, so that it was the contents rather than the tone of the briefing that Meron found inflammatory. We had a pleasant lunch with Meron, and a rather low-key working meeting with Eini and his team. We also briefed the IAI people, including Bully Blumkine and IAI board chairman David Ivri, who was slated to move to the ministry of defense on June 1. Blumkine somehow managed to keep his temper under control—no small feat for him.

Thus far there had been no big explosions, and I felt relatively relaxed by day's end. I was free that evening to visit the Wailing Wall, something I did at least once on every visit to Israel. I then dined with my friend David Rosen and his lovely family. David was a brilliant young rabbi who had served as chief rabbi of Ireland in his early thirties and was now heading a major institute devoted to bridging the gap between religious and nonreligious Jews. David has since moved on to even more formidable challenges: he is perhaps the Israeli Orthodoxy's leading exponent of improved interfaith relations, and has developed close ties with the various Christian and Muslim hierarchies that flourish in the Middle East.

I returned to Tel Aviv to brief Rabin the following morning. I recognized that our entire effort turned on his reaction to our work. Rabin had already placed the Lavi project under tremendous pressure: in April 1985, just after we briefed our preliminary findings to the Israelis, he had ordered an annual ceiling of $550 million—in 1984 prices—on expenditures for the Lavi. Israeli supporters of the Lavi said that the $550 million ceiling was an average over the life of the project, allowing for some years in which spending might exceed the limit. Nevertheless, American inflation, then running at about 3 percent, was certain to squeeze the project ever more tightly with every passing year. In any event, our estimates, which postulated expenditures of $900 million or more in a few years' time, rendered the project impossible under Rabin's stated terms. Interestingly, Rabin himself was rather sphinxlike about just what it was he meant by the ceiling he had promulgated.

We spent well over an hour with the minister. When we met first for a quick private chat I found Rabin easier to speak to than before. We talked a bit about Oxford and Richard Crossman, the former Labour minister who was both a Judeophile and very pro-Israel. I had heard much about Crossman but had never met him; Rabin knew him well and reminisced about him in the warmest of terms.

Just as I was about to start my formal briefing in the conference room, the tea lady appeared at the door, combat boots and all. Rabin took the course of preemptive surrender. Taking in one of her Medusa glares, he held his head in his hands as she slowly, deliberately, handed out tea to the twenty-odd people in the room. After that, the briefing was a breeze; at least Rabin seemed to take the results very much in stride. Tom Pickering, who had accompanied us, was extremely pleased by the outcome.

We then traveled to Jerusalem to brief a very sympathetic finance ministry staff. I was surprised that Yitzchak Modai, the minister, had not appeared for the briefing; it was a foretaste of things to come. The mercurial Modai ultimately would side with the Arens faction as the Lavi's fate hung in the balance.

That evening Eini hosted us at his home. Although not at all religious himself, Eini knew that I kept kosher. Israelis serve meat at several points during a meal, and on the first such occasion, I asked the waiter, who seemed a friendly enough fellow, whether the meat was kosher. He told me it was not. I asked whether any of the meat that we would be served was kosher, and again, he replied in the negative. I was both surprised and offended that Eini was serving us only non-kosher meat. Considering that we were in Israel, where kosher meat was not exactly hard to find, I figured he must have gone to some trouble to find meat that I couldn't touch. The waiter understood my predicament and managed to keep me well fed without my having to violate my dietary practices. As for Eini, I suppose he got a good laugh out of it all.

The trip to Israel was a very short one, and had been far quieter than I had anticipated. I did not expect things to remain quiet for long, and was proved correct once I returned home.

The Lavi's Israeli supporters didn't wait long to counterattack. Within two days of our return home, *Aerospace Daily* was carrying comments from "Israeli sources" (I assumed that meant Eini) that our high production cost estimates were due to "inadvertent errors" in the way we totaled our data. These sources also alleged that "'the Zakheim report' is politically motivated." Said one, "If Israel builds this aircraft at the cost it claims, then how can the Air Force justify the prices it pays for its planes?"[3]

The Lavi's supporters simultaneously opened a second front in Israel, with "a senior defense source" telling the *Jerusalem Post*'s Hirsh Goodman that "a 10 percent mistake . . . has already been discovered, and we have only begun to scratch the surface."[4] The front-page story stated the assertion of "Israeli officials" that the U.S. findings had come "as no surprise." A senior official went on to say that some of the discrepancies were due to "the different way in which Israeli and American industry operates, and to the application of American costing procedures and criteria to an Israeli situation."

The Israeli no doubt was referring to the Israeli contention that our estimate of Israeli hourly labor rates was twice as high as it should have been ($44 an hour instead of the Israeli estimate of $22 an hour). Both sides had been aware of this difference for some time. We had asked for the Israeli "wraparound" rates, and the Israelis had been unwilling, and claimed that they were unable, to provide them. Accordingly, we applied the best analogous estimates we could.

Our discussions with the Israelis and our visits to Israel had led us to conclude that there really were no special "Israeli" ways to build a plane, and that the cost categories were no different in Lod from what they were in Fort Worth or St. Louis, where American fighters were built. The argument that Israel was somehow different was to be a recurring theme of the Lavi's supporters, however, especially once we began to examine alternatives. Though it was quite true that there had never been an Israeli minister of defense as enamored of systems analysis as Robert McNamara, who introduced the discipline to the Pentagon, it was also true that Israel's "best and brightest" were trained at institutions such as Harvard's Kennedy School, where systems analysis had been *de rigueur* for two decades. For our part, all we could do was insist that our estimates be judged on their merits, and nothing more; in any event, the difference in labor hours did not, in and of itself, account for the disparity between the estimates.

The Israelis also contended that we had overestimated the cost of the Lavi's new engine. The preliminary agreement between Israel Aircraft Industries and Pratt & Whitney, the engine's American manufacturer, called for a price 50 percent lower than our estimate. Of course, we knew about the contract—the Israelis had never been shy about pointing out our error. But we noted, to deaf Israeli ears, that the preliminary contract was just that—preliminary. It was only for the first thirty engines, whereas the Israeli program called for ten times as many engines and had originally postulated that these engines would be built in Israel. We knew, and the Israelis did as well, that there was no firm contract for any but the first nine engine prototypes. Moreover, there was virtually no likelihood that the Israelis would ever build the engine, because the Bet Shemesh plant could not even coproduce those engine parts that Pratt & Whitney had assigned to it. We therefore significantly increased engine costs. The Israelis simply chose to ignore the reality of the failed built-in-Israel engine program.

The Israelis even tried to have the argument both ways. After having prevented us from calculating the program's cost on the basis of real engineering estimates, thereby forcing us to use what are termed parametric methods, the Israelis criticized us for using the wrong abstract approaches. In one instance, they argued that we should have employed a methodology developed by the RAND Corporation, the government in-house think tank that provided major support work for the air force. Not surprisingly, to us, at least, that particular approach would have understated the Lavi's costs; and I told the Israelis as much in a letter to Zvi Tropp.[5]

Of course, the technical differences were unlikely to sway the general Israeli public in favor of the Lavi. So the unnamed Israeli officials who provided backgrounders to Hirsh Goodman and other journalists played on the public's strong distrust of Caspar Weinberger and "his" Defense Department. Goodman's unnamed Israeli officials told him that the study could not be objective anyway, since it had been carried out under the auspices of the Pentagon, which opposed the project.

There was no comeback to the last argument. As to the others, in addition to citing our own technical position, I began to repeat the baseball mantra to anyone who would listen: "I call 'em as I see 'em." But then, a fan's usual response to that remark is "Kill the umpire," so I didn't expect to get much relief from our critics.

The Israeli criticism continued, of course. Nevertheless, the more thoughtful of the Lavi's supporters recognized that they no longer could proceed with their project away from the public spotlight, and that they would have to provide a far stronger justification for the project if it was to proceed as planned. As one of them stated to Ze'ev Schiff, one of Israel's leading defense reporters, "There are two possibilities: Either the research of the American experts is mistaken regarding several basic issues, or if they are correct, we are facing a difficult situation, which requires us to make new decisions." Schiff went on to observe, "These statements show that last week's visit of the Pentagon delegation to Israel constituted a turning point for the Lavi project."

Schiff's analysis reiterated what we were insisting: the United States was not telling Israel what to do; that was a decision for the Israelis themselves, and, he stated, "it is clear that the argument over the Lavi will deepen."[6] That, in fact, was exactly our intention. We had no desire to say much more about the Lavi at this time; the report was now in Israel's hands. Our strategy was to let the Israelis stew in their own analytical juices, while we pursued the cost analysis of the new Israeli naval program. We created a senior-level "Dolphin/Saar Executive Steering Group" along the lines of the Lavi Steering Group to oversee the course of the study. The group consisted of many of the same people as the Lavi group—representatives of ISA, State, OMB, and the NSC, as well as the offices of the secretary of the navy and the chief of naval operations. It was this group that formally approved the study's terms of reference and its various drafts. By functioning as smoothly as it did, it underscored for the Israelis the degree to which the various agencies of government were united on their approach to Israeli budget issues, and therefore also reinforced the message that we tirelessly reemphasized to the Lavi program office: like the naval modernization study, the Lavi effort represented the views of the entire administration, not just those of one individual or one department. It was a message that took the Israelis nearly a year to digest.

The Israelis were not about to allow me or my staff to devote ourselves entirely to the analysis of the Israeli navy's requirements. Within two weeks of our trip to Israel, my deputy, Dick Smull, was visited by representatives of the New York purchasing office and the MoD in Tel Aviv, to check on the processing of Lavi contracts. This was soon to become a major issue as I exercised more and more of a hold on those contracts, but at the time we were letting them through at the rate of about one a week. The visiting Israelis also passed along a detailed list of questions that Menachem Eini had about the report. It was interesting that Eini, who had done so much to obstruct our getting cost information for our study, now was asking very precise technical questions about *our* cost estimates. In fact, the list was breathtaking for its chutzpah, given Eini's own behavior. The queries included:

- a request for items included in the airframe cost analysis;
- specific definitions and price details of materials in different definitional categories;
- a detailed list and pricing of subcontracts;
- the basis for our system-engineering and program-management estimates;
- the specific cases for which engineering change orders were assumed;
- the elements included in nonrecurring costs for production; and
- a request for a breakdown of the costs of an exhaustive list of categories including, but not limited to, airframe, avionics, engines, system test, support, and operations and support.

Dick passed this laundry list along to the air force's Linda Hardy; she did her best to demonstrate exactly how we had reached our conclusions.

Linda's analytical work by no means satisfied Menachem Eini. He was back to see me in mid-March, and following discussions with him about our life-cycle cost estimates, fired off a new four-page questionnaire addressing whatever aspects of our estimates he had missed the first time around. His March 25 letter accompanying the questionnaire stated that the Israelis had decided to send a team of six people from the MoD and IAI to the States to "study in depth your recent report, to understand its database, and to try to reach an agreement with your people on cost estimates." In addition, he requested a briefing regarding the report, its methodology, and its database. Finally, he asked that we provide as many answers as possible to his questions prior to his team's departure from Israel.[7]

I forwarded Eini's letter to the comptroller of the air force, noting that "it is vital that we and the Israelis reach agreement on the cost estimate of this program," and requested Linda Hardy by name.[8] I then informed Eini that we were prepared to meet with his cost analysts in mid-April. I also enclosed a set of detailed answers to his first questions. These included our methodology for calculating wraparound rates, avionics costs, engine costs, and labor hours.[9] We knew that no matter what we showed Eini, he would be unstinting in his criticism unless we actually accepted his estimates and repudiated our own. Nevertheless, I preferred that he criticize us for our analysis rather than turn tables on us and say that we were withholding information from him.

During the April 14–16 meetings the Israelis pressed us again for more details on the assumptions that underlay our study. By now we were certain that they were going to issue a report of their own. We also did not doubt that it would be an "evaluation" of our study. Given the tenor of the questions they had put to us, let alone the unceasing leaks and disparaging remarks about our effort, we were under no illusions as to what it would say. In any event, it had in effect already been previewed in the Israeli press even before it was written.

Sure enough, the Israelis formally counterattacked in late April. I received a copy of their "study" in a note dated April 29, 1986, which summarized the Israeli team's findings. The note included the Israelis' reaction to our most recent round of talks two weeks earlier as well as to our replies to their written questions. Their "careful review," as they called it, was meant "to ascertain the basis of the substantial gap" between their estimate of the Lavi production unit cost and the projection of the DoD assessment. They asserted that there were "very basic misunderstandings" in the economic assumptions applied by DoD in its analysis. And they insisted that "after a careful review" of the document and receipt of our explanations they had "discovered that in fact there are fundamental errors in the DoD assessment," based on an "understandable" confusion regarding the "substantial" differences between the pricing/cost structures used by Israeli industry and those used by U.S. industry. The DoD conclusions, they asserted, were not based on an analysis of the industries in Israel, but rather on direct analogies to comparable activities by U.S. industry.

There was certainly an element of truth to the last assertion, which underlay their repetition of issues that had already been leaked *ad nauseam* in the Israeli and American trade press. Of course we had used "analogies"; the Israelis had provided us with little more to work with. Our team had visited Israel's industries and had concluded that they conducted business very much as ours did. A fighter was a fighter was a fighter, and building one involved similar principles and costs in different countries. There really were no shortcuts. Such differences as existed with respect to wage rates did not fundamentally alter our estimates, which were based on many other factors—such as overhead costs—that the Israelis continued to refuse to address. In any event, after recycling their critiques one more time, the Israelis presented us with "corrected" costs based on the meetings that had taken place in mid-April. Their total reduction amounted to a unit cost that declined by about $7 million, bringing the revised DoD figure in line with their own. This became the new Israeli rallying cry, one which was picked up by their supporters in the United States as spring turned into summer.

7

Rabin Blesses a
Second Lavi Report

Throughout the late winter and spring of 1986, even as we beavered away on the naval modernization study and initiated our planning for the alternatives study, I had other tasks to carry out and other trips that they engendered: to Paris and Worcester, England, for meetings of the Four Power Armaments Directors (from the United States, France, Britain, and Germany), to London and Bonn for talks with the federal ministry of defense. I was involved in a host of programmatic decisions, a major one being the future of the division air defense gun, an army program that excelled only in the amount of money it continued to waste. I enjoyed the work, but found that the travel was no longer exhilarating as it once had been. I began to feel that I was away from my family far too frequently, and, though I could sleep on airplanes and didn't suffer from jet lag per se, I nevertheless found myself constantly tired.

I retain few memories of these trips. One incident that does stand out in my mind was my leisurely walk around Worcester Cathedral in late March with Dr. Karl Schnell, the German armaments director, a pleasant man about twenty years my senior. As we walked out of the cathedral, admiring its light and soaring architecture and its surrounding greenery, Schnell suddenly began to tell me of his youth. He had served in the Luftwaffe during World War II but, he added, he had stood up to the regime. Responding to my manifest curiosity, Schnell related that he had been a student at a school run by monks—I cannot remember which order, though I do recall that it was not Trappist or any other with which I was familiar. When he was thirteen the Nazis closed down the school. He transferred to another school and there, as a teenager, joined the Hitler Air Youth Corps. On initiation Sunday, he said, the cadets were forbidden to go to church. He was outraged, and, after his exercises, he attended assembly in civilian clothes, rather than in his Hitler Air Youth Corps uniform. He then went on to say, with pride, that this act of defiance got him into serious trouble, though he did not say just what kind of trouble it was.

I really didn't know how to respond. It was not the first time that a German had, in effect, left me speechless. I recall the first German I had ever met, on a railway platform in Zurich during my junior year at college. All I had known of Germans was that the Nazis had murdered my father's parents and his two sisters, and that many of my friends' parents had tattoos on their arms. I didn't know that the young man I was addressing was a German. We simply began to chat, and I asked him where he was from. "I'm a German," was the reply. I froze, and blurted out, "I'm a Jew." He stared at me for a second, and said, "It's okay. My father was a socialist; he spent lots of time in Dachau." His straightforward demeanor broke the spell; ever since then my attitude to Germans, as to all people, has been that they are innocent until proved guilty—even if they are of an age when they might well have served in the Wehrmacht, or the SS.

I liked Schnell. But his remarks threw me. My immediate reaction was to wonder at just how peculiar the Germans were. How ludicrous it was for Schnell to be so proud of his schoolboy act of rebellion and to attempt to give it cosmic proportions as some sort of renunciation of the Nazi regime. Over time, however, I came to realize that, from his vantage point, he *had* in some sense resisted the Nazis. He had demonstrated that he was not merely one of the herd; he had risked his future to make a statement.

As we walked back to our hotel, my mind wandered off to speculation about what Schnell had done after he left the Youth Air Corps. I decided not to pursue that line of thought. Schnell had felt he had to say something to me, and he did. It was something of an apology on his part, an apology for what his people had done to mine. It was not the first time a German had, in effect, apologized to me for Hitler and the Holocaust. But I found that I most welcomed the apologies when they were not really apologies at all; the young man whom I had met that day in Zurich didn't really apologize—he didn't have to.

Schnell and I also talked about the Lavi, about which he was noncommittal. Few of the Europeans I met had much sympathy for the Israelis, and many hoped that the program would fail. The Germans were, however, a leading contender for the submarine program. Their Hanseatic shipyards were hurting; business was down in Emden and Kiel, and unemployment, the bane of German political economy, was on the rise. The Germans desperately needed an Israeli order, and were anxious that the Pentagon support it. They were therefore not about to alienate the Israelis by criticizing the Lavi, though they had no incentive to support it if that meant jeopardizing the Israeli navy's program.

However much I traveled, or, for that matter, worked on issues other than the Lavi, the aircraft remained very much a central part of my life. I continued to discuss the project with visiting Israeli officials, including Prime Minister Shimon Peres when he came to Washington in March 1986. I also continued to address Jewish audiences on U.S.–Israeli relations, especially the future of the Lavi.

More important than speeches or meetings, however, was the question of what I intended to do next. The Israelis assumed that I would not rest merely with a report that argued that their cost estimates were too low. My superiors at the Pentagon insisted that I not stop with that report. Weinberger and Iklé were even more certain, if that was still possible, that the Lavi was a waste of money. We had to move ahead to ensure that Rabin would kill the program.

By late April we were already immersing ourselves in the data that compared the Lavi to our most modern fighters. We focused on the air force F-15, probably the world's leading fighter; the F-16, a multirole fighter that, like the F-15, was already in the Israeli inventory; the F-18, a new multirole navy fighter/attack plane; and the AV-8B Harrier, a relatively short-range attack plane that could take off and land vertically.

I had no way of knowing how the Israelis would respond to our new effort. I hoped to win Defense Minister Rabin's support for what we were doing. That, however, had to wait until my next trip to Israel, ostensibly to review our progress on the naval modernization study. We had to begin our fact-finding for the second Lavi study with some care. We did not want the Israelis to make a premature fuss over our effort before I spoke to Rabin. Accordingly, even our team meetings were more discussions of methodology and tactics rather than task-oriented exercises.

At the same time, work proceeded apace on the naval modern-ization study. I felt strongly that Israel needed to replace its Gal subs, the Joint Staff's report notwithstanding. I was sensitive to the danger posed by the Gal boats' structural problems: I could not forget that Israel had already lost the *Dakar* at sea in 1968; with such a small navy, it could not afford to lose another boat because of structural failures. In addition, I did not have problems with the proposed range of the boat, which far exceeded that of the Gal class, even if it was meant for long-range special operations, possibly against Libya. Some analysts wrote that the boats were meant to be strategic nuclear deliv-ery vehicles. I considered that theory to be outlandish; the boat's design simply didn't allow for such a mission.

While I therefore sympathized with the Israeli goal of obtaining new subs, I did nothing to encourage their increasingly faint hopes of acquiring submarines from the United States. I personally had long supported U.S. construction of diesel boats; our shipyards were dying from lack of work, and I was certain that we could hold our own in the burgeoning world market for diesel submarines. But my friend John Lehman was as adamantly opposed to the idea as any submariner. He was enmeshed in so many other battles with his uniformed officers that he simply could not take on another, how-ever strongly he sympathized with Israel generally and with its navy in particular. Since I in any event felt that the Israelis should not waste their energies in a losing battle against the nuclear mafia, John's reticence clinched the matter for me.

I also did not sympathize with a second Israeli program objec-tive, the construction of surface ships in Israel. The Saar corvette was an entirely new Israeli design that was likely to be attractive to many Third World states. It displaced about twelve hundred tons, nearly twice as much as the next-largest Israeli missile boat, and was packed with the latest in weaponry and electronics. It was natural for the Israelis to want to employ American funds to build the corvette in Israel. But we needed work too. Assistant Secretary Rich Armitage and I—as well as Lehman—felt strongly that the missile boats had to be built in the United States. The Israelis, to their credit, agreed to postulate the program's costs on the assumption that the subs would initially be built in Europe and the boats in the United States. As a result, both sides were able to hunker down to the business of finding the most reasonable estimate for the program without the fireworks that the Lavi effort produced.

My opportunity to earn Defense Minister Rabin's blessing for a follow-on Lavi study presented itself earlier than I had anticipated. Just two weeks after Passover, which fell during late April in 1986, Rabin made his semiannual pilgrimage to Washington. His trip followed hard on the heels of the formal Israeli evaluation of our Lavi study, and we therefore really had no way of knowing how he would respond to a suggestion for yet another study—a third in the space of less than two years—to conduct the dreaded evaluation of alternatives to the Lavi.

On Iklé's instructions, I organized a full-dress Lavi briefing for Weinberger on Monday, May 5, in anticipation of his scheduled meeting with Rabin the following day. Despite all the publicity surrounding the Lavi controversy, Weinberger had never been fully briefed on the matter. Instead, he issued his instructions on the basis of initialing one-page decision memos, effectively leaving Iklé, and by extension me, to manage the American effort to kill the program.

Iklé attended the briefing, as did Don Hicks, who had recently become under secretary for research and engineering. I had known Hicks since my days at the Congressional Budget Office, when Don was a senior vice president of Northrop. A highly intelligent and voluble man, Hicks made no secret of his dislike for the Lavi. Since Don was a strong supporter of Israel, I suspected that his attitude was shaped in part by his involvement in Northrop's ill-fated effort to sell the F-20 aircraft to the Israelis. Whatever his motivations, Hicks was a valuable ally, assuring that there was unanimity in the department in support of my efforts—a rare experience for anyone working in the Office of the Secretary of Defense.

Fred Iklé had asked me to lead the conversation, and I stressed that our findings clearly indicated that the Lavi was an overpriced program. Weinberger, ever suspicious of Israeli intentions, nodded frequently in agreement. He made it clear that he would strongly support our findings in his meetings with Rabin, including their "one-on-one" meeting, which was attended by the principals and their senior military assistants.

As matters turned out, Weinberger was as blunt with Rabin as the latter was wont to be with his own interlocutors. While the plenary meeting and formal lunch, which I attended, were pleasant enough—Rabin agreed to join the American effort to develop strategic defenses—Cap stunned the Israeli in their private session by telling him in no uncertain terms that he should go back to the cabinet and cancel the Lavi. Since I was scheduled to meet privately with Rabin at 5:00 that afternoon, I expected to bear the brunt of his reaction when we got together at his suite in the Grand Hotel.

As I emerged from the elevator to the floor that Rabin had taken over at the hotel, I was immediately encircled by several Israeli security personnel. Just as our own Secret Service people are easily identified by the "hearing aids" with which they communicate, Israeli security types always seem to have heads shaved in U.S. Marine Corps style. I exchanged a few words of Hebrew with them and was ushered into Rabin's lounge. I expected him to be surrounded by his aides, notably Colonel Shimon Hefetz, his military aide (now General Hefetz, military aide to the president of Israel); and the rotund Eitan Haber, who has divided his life between journalism and serving as Rabin's spokesman. But only the army colonel sat in on the meeting.

My apprehension about the meeting was not unjustified. Rabin looked tired and did not appear to be in a very good mood. I had rarely exchanged preliminary pleasantries with the defense minister, and today was no different. He came straight to the point, telling me that the United States was out of line in holding up Lavi contracts. I responded ambiguously. In fact, I was deliberately slowing down the American support process by insisting that all Israeli requests for DSAA approval of American support for the Lavi be routed to me before any further action was taken. Once my office received the Israeli requests, we sat on them for at least a week, taking whatever time was needed to ensure that we were not passing on to the Israelis any technology that did not fall within a strict interpretation of our support agreements. Our action prompted irate calls from Abraham Ben-Joseph at the New York purchasing mission, and I had expected a formal Israeli protest when Rabin came to town.

What surprised me was that Rabin quickly followed up his protest by stating that he was capping the Lavi at $550 million for the next fifteen years. This represented far more budgetary pressure on the Lavi than had previously been the case, since a fifteen-year cap guaranteed delays, a low production run, higher unit prices, and, ultimately, more pressure for cancellation. Encouraged by his statement,

I raised the matter over which I had requested the meeting. I asked Rabin whether he would object to our evaluating potential alternatives to the Lavi. Rabin replied, indeed stressed, that he was not asking us for any alternatives. At the same time, he said with what seemed to be a small twinkle in his eye, we could of course suggest them. After all, we were friends and allies, and friends always were in a position to offer suggestions.

As far as I was concerned, that was all I needed. Rabin had opened the next set of locks on the Lavi canal. Once again, at a critical juncture, he had not prevented us from pressing forward with an effort that could bring the program crashing to a halt. I left the hotel admiring the man's brilliant combination of tactical caginess and strategic foresight, a combination that continued to stand him in good stead as he wove his way through the thickets of the Middle East peace process, earning the Nobel Peace Prize along the way.

Although I had spent most of my time in the months before my meeting with Defense Minister Rabin consumed by matters other than the Lavi, the Israeli press, and to a lesser extent the American Jewish press, turned the heat on the political leadership in Jerusalem. The Lavi was now a front-burner issue, possibly Israel's leading issue; and the debate was not limited to the question of direct costs, but increasingly addressed the viability of the program in the larger context of Israeli security.

One example of this new degree of scrutiny was a rather long article that appeared in the Labor daily *Davar*'s weekend supplement in early April. The article bluntly characterized the Lavi as "a gamble," and "a dead weight around our necks."[1] The issue, stated the paper, was not the plane's capability; it noted that even the Pentagon had acknowledged its technological and performance improvements. Rather what was at stake was the health of the economy in general, and of the defense budget in particular. Reflecting the grumbling that was getting louder within military circles, the article noted that both the army's Merkava tank and the navy's modernization program would be "damaged."

The *Davar* analysis put the case exactly as we hoped it would be carried to the Israeli public. Moreover, it underscored three aspects of the issue that had hitherto not received much publicity in Israel, but were crucial to any case against the plane. The first was the underlying and very implicit assumption that governed the strategy of the Lavi's proponents, most notably Moshe Arens. As *Davar* put it, "it looks as if the Israeli government is harboring the hope that the American Administration will not only finance the plane's development, but its production as well." The second was the fact that all Lavi advocates tended to overlook as they wrapped the plane in nationalistic blue and white: that "the Lavi is not a homegrown item from the tip of its beak to the edge of its tail. The American component in its development comprises at least 50 percent."

Finally, and crucially, the article highlighted the still-little-noticed change in the nature of congressional support for the Lavi. It recognized that for the first time, the Foreign Assistance Act for fiscal year 1986 permitted Israel to apply offshore procurement funds to projects *other than the Lavi*. "The channeling of these funds into other programs," noted the article, "would probably make a more significant contribution to Israel's security than the Lavi." That precisely was the case we intended to make as we pursued our new alternatives study.

Faced with the need to defend the suddenly beleaguered plane, the Lavi's proponents often were so carried away in their attempts to make the Lavi a symbol of Israeli jingoism that they resorted to arguments that seemed to make Americans look like fools. Thus, for example, IAI chairman David Ivri, who was soon to replace Mendy Meron as the ministry of defense's top civil servant, told the left-wing *Al Hamishmar*, which was generally critical of the project, that the Lavi was actually a "national *American project.*" He added that the choice was "the purchase of an Israeli plane with American money or an American plane with American money. This choice boils down to providing employment for U.S. industry or for its Israeli counterpart, and the preferred alternative is to provide jobs for Israeli workers."[2] Needless to say, the Ivri statement infuriated us in the Pentagon. It was bad enough that we felt we were being taken for suckers; it was worse that Ivri was advertising the fact.

The irony was that it was not really a matter of American jobs versus Israeli jobs, however noxious and *chutzpahdik* such argument seemed to be. Rather it was a case of American jobs versus other American jobs, with Israel Aircraft Industries getting a lot but not all of the business if the Lavi was procured, and only somewhat less if Israel chose to acquire an American plane.

Ivri was not loath to speak to the press about our analysis, even as his transfer to the MoD became imminent. Late in April, he told the

Israeli air force journal that our estimates had been dropping steadily, from an initial guess of $10 billion for research and development costs to the $2.6 billion that he claimed was in the report. Since IAI had stated that research and development (r&d) would total $2.2 billion, he argued that it was the DoD whose estimates were being refined in the direction of those of Israel, and that, once other errors were accounted for, the Israeli estimates would be validated.

Ivri was being disingenuous. We had made clear to him, as to all others whom we had briefed in April 1985, that our initial estimates were based on very little in the way of information from Israel. Indeed, we had stated that it was for that very reason that we needed better input from Israel, because we were certain that we had over-estimated our costs. Moreover, the $10 billion figure was only the U.S. Air Force projection. Even in April 1985, the Office of Program Analysis and Evaluation put the estimate at $3 billion. As for the Israelis themselves, their r&d estimates had grown from $1.5 billion to $2.5 billion, and subsequently were revised downward to $2.3 billion. Moreover, our report had indicated that the differences between us and the Israelis centered on production costs, for which our estimates were over 50 percent greater, as opposed to r&d costs, for which the difference was less than 15 percent. Ivri knew this too. He knew that we differed over wage rates—our estimate put Israeli wages at over twice the IAI estimate—but that a good part of the difference could be explained by the related overhead costs that IAI and the government did not furnish.

Naturally, for Ivri, when he served as chairman of IAI, jobs at his company was his highest priority. He was a partisan, and had every right to take a partisan approach to disputes over costs, schedules, or anything else that might have prejudiced his program. He did not change his attitude once he reentered the ministry of defense, however, even though his own generals, including air force generals, were pressing him not to push ahead with the project. Indeed, for a time Ivri served as *both* MoD director general *and* chairman of IAI, which is a government-owned corporation. Such a dual role would be considered a conflict of interest in the United States. The fact that it was acceptable in Israel not only underlined the very different way business was done there but also explained why the MoD was so willing to accept IAI's figures. In a very real way, and to a greater extent than most Americans realized, IAI and the MoD were one and the same.

Ivri, incidentally, remains director general of the ministry of defense. He heads two of the groups that are the cornerstone of the special American-Israeli relationship: the Joint Political Military Group, which addresses planning issues, and the Joint Security Assistance Planning Group, which programs priorities for future American security assistance to Israel. Ivri has never repudiated his statements about America that he made during the Lavi episode, including the one about getting the United States to fund Israeli jobs at the expense of the jobs available to its own workforce.

Even as the Israelis were taking us to task for our cost figures, a new actor had emerged on the scene, promising to furnish yet another set of Lavi estimates. Early in March, Lee Hamilton, the chairman of the House Foreign Affairs Subcommittee on Europe and the Middle East, had asked the General Accounting Office, the audit agency for Congress, to compare our cost estimates with those of the Israelis. We received formal GAO notification of the study on March 20, indicating that the study effort would commence on the 24th. The notification letter, addressed to Cap Weinberger, stated that the GAO intended to undertake a sweeping review of the program. Specifically, we were told that the GAO would examine:

—the history of the Lavi program, including how and why the aircraft design requirements changed since its origin;
—an estimate of the U.S. security assistance and Israeli funds expended on the program and how and to whom those U.S. funds were dispersed;
—a comparison of U.S. and Israel's cost projections, noting any substantial differences, and GAO's assessment of their reasonableness;
—implications which the cost estimate could have on future Israeli security assistance requirements;
—expected production of the Lavi, including expectations for export sales; and
—the extent to which funds for the Lavi impact on the adequacy of funds available for other Israeli military requirements.

The letter went on to say that the GAO intended to conduct its research at State and Defense, and left open the possibility of meetings with Israeli representatives in Washington and New York and even abroad, presumably in Israel.[3]

Clearly, the GAO was setting itself up as a referee between ourselves and the Israelis, but we were not overly concerned about that development at all. Hamilton was, and is, one of most respected members of Congress. Some Israelis and staffers in the American Jewish representative offices that dot the Washington landscape did not warm to him because, although he supported Israel, he was not a knee-jerk advocate, like some others in the Congress. But it was precisely his judicious attitude to matters relating to Israel and to other issues that had won him so much respect around town and internationally.

When I had met with Hamilton, it was clear that he was not enthusiastic about the Lavi, but was not ready to oppose it. In this regard he shared a common view with David Obey, the chairman of the House Appropriations Foreign Operations Subcommittee, who made it clear that he simply did not have the votes to make a fuss over the project. Hamilton's request to the GAO and the agency's terms of reference for its study were thus very much in keeping with his balanced but mildly skeptical approach to the program.

Despite Hamilton's well-deserved reputation, we did retain some residual apprehension about the study. To begin with, the GAO's work actually has an uneven reputation among defense budget analysts: some of its studies are first-rate, others are quite pedestrian. In addition, the DoD has often clashed with the GAO. The Defense Department will argue that the GAO's findings have been overtaken by events, while the GAO will, ever so politely and indirectly, accuse the department of dissimulation. I had interacted with GAO analysts while working both at CBO and at the Pentagon, and my experiences with them likewise had been quite mixed. I had heard that the analysts they assigned to the Lavi job, particularly Al Huntington, were tough-minded no-nonsense types, and I hoped that would be the case; dealing with the Israelis was not going to be easy.

We knew that the Israelis were seeking yet another means of countering our study that went beyond mere attacks in the press, and we learned that they had hit upon the GAO as an appropriate counterweight. We were under no illusions about the Israelis' intentions or methods: they would exert all the political pressure available to them to ensure that the GAO report came out with estimates that supported their own.

I worried that the Israelis might actually succeed in getting the numbers they wanted. I had found in the past and was again to find in the course of the GAO's study that the institution was not entirely immune to political pressures. Staff analysts could find themselves having to modify their conclusions at the behest of their higher-ups, who themselves were unhappy about the pressures being placed on them to come up with the "right" conclusions. To the GAO's credit, it usually manages to find a way to assuage those who are applying the pressure while nevertheless preserving its intellectual honesty—no small feat indeed. With an issue as highly charged as Israel, particularly given the GAO's record in calling straight—and to the Israelis, often unpleasant—shots about the security assistance program, it was clear that the pressures, or at least the specter of pressure, would haunt the analysts who had been assigned the investigation. We could only hope that Huntington and his colleagues would stick to the numbers and avoid the politics that surrounded them.

Even as Ivri and his colleagues were pressing the case for the Lavi and the GAO was gearing up for its own study, we for our part had begun to lay the groundwork for the larger debate that was reflected in the *Davar* article of early April. Perhaps our most widely reported effort was my critique of Israeli management that I presented in an address I had given to the American Jewish Congress on American-Israeli relations.

The speech highlighted my concern that the MoD was providing insufficient oversight of major Israeli programs, including, of course, the Lavi. I urged Israel to tighten its military planning and budgeting system so that it could avoid large-scale mistakes that it simply could not afford. I was at pains to emphasize that while my counterparts in Israel were world-class in their expertise, nevertheless there simply were too few of them to manage the array of projects that the defense ministry had undertaken. As a result, the officials who did manage these programs had insufficient time to devote to critical program, system, and cost-evaluation tasks. My concern, put bluntly, was that "these are matters that simply cannot be left to contractors. Their job is to produce a product, not to critique themselves."[4] I added that in the past, the Israelis had managed to avoid the consequences of a failure to manage programs more carefully because so much of their equipment had been procured from abroad. As a result, critical supervision over the research and development phases of each system had been provided by the vendor states. As Israel came to produce more and more of its own systems, however, what I called "secondhand analysis" was simply not enough.

Much of what I told my audience, which was reported in both the international defense press and in stories filed by Wolf Blitzer, whose articles appeared in the *Jerusalem Post* and in virtually every other English-language Jewish newspaper worldwide, derived from my long, confidential conversations with MoD economist Zvi Tropp. I could not get over the fact that Tropp had a staff of only three people to cover the gamut of actual program management. And I had ten times as many just to review the one Lavi activity!

Nearly a decade has passed since those conversations with Zvi, and I am not entirely sure that Israel has added many people to its program review staffs. The number of home-grown Israeli projects keeps rising, however, so that the requirement for such supervision has not disappeared. It is an issue that I have often raised with my Israeli colleagues since my departure from the Pentagon, one that will not go away as long as Israel continues to produce weapons for its own military and those of other states.

In addition to my observations to the American Jewish Congress about cost control, I raised another issue that we had tried to impress upon the Israelis, initially with very little success. This was the matter of American budget constraints. Although the U.S. Congress had passed the Gramm-Hollings-Rudman budget legislation of 1985, Israel seemed not to notice. This act had fostered a completely new congressional attitude to spending and signaled the beginning of a real decline in defense expenditures that continues to this day. Under the terms of the act, if congressionally preset spending limits were breached, rescissions would result. Such cutbacks would also mean reductions in security assistance for the year in which the spending limits were exceeded.

I warned that the new congressional atmosphere meant that Israel could not expect ever increasing levels of security assistance. In fact, it was in order to impress upon the Israelis the seriousness of this argument that I normally was accompanied to Israel by Len Zuza, OMB's senior officer for the Middle East. At every opportunity I asked Len to explain the overall budget atmosphere to skeptical Israeli officials, and often asked the State Department representative to offer his own concurring views.

The Israelis didn't pay attention, of course. They clearly believed that their case was different, that somehow the Congress would find a way to keep pouring more money into their coffers. They didn't even take to heart the fact that as a result of the new legislation, when fiscal year 1986 funding limits were broken Israel had been forced to return $76 million to the U.S. Treasury. Certainly, when they were informed that they would have to sign a check to the U.S. Treasury, the Israelis were shocked. They tried to get AIPAC to work out some arrangement whereby they would be treated as a special exception. AIPAC undertook a valiant but doomed effort to get the Israelis special treatment. Israel had to pay up.

Still, the Israeli attitude did not change even then. Much like Pharaoh, who refused to face the evidence of the various plagues visited upon his people, many in Israeli officialdom, and foremost among them Moshe Arens, remained wedded to the belief that somehow there would be enough American money to finance the Lavi. But we were not trying to convince Arens. To my mind, he was a lost cause. I was trying to get the attention of the Israeli public and of the American Jewish public that resonated to Israel's needs. The change in attitude that I sought when speaking to the American Jewish Congress did not come quickly, but it did come. The shock of that $76 million check that Israel paid out had some effect on the American Jewish leadership; my guess is that it also had an effect on Yitzchak Rabin.

8

The Lavi Gets
Personal

The Lavi episode had one unintended consequence for me: I had become "newsworthy." I found myself in constant demand to speak to Jewish organizations, to the Jewish press, and, more generally, to journalists who smelled a good story. I had always had a pretty good relationship with reporters. During my first years at the Congressional Budget Office, which also were the first years of the CBO's existence, we were constantly calling press conferences to draw attention to our work and finding that very few reporters showed up. We received very little publicity for our analytical work in the mid-seventies. By the time I left CBO, however, it had developed more of a public profile. We were contacted more frequently by reporters, and I had developed some good working relationships with a few of them.

Coming to the Pentagon, however, I was unsure whether the rules that had stood me in good stead on the Hill—essentially to assume unless proved otherwise that a reporter was honest and would play by the attribution rules—were applicable to my new job at the Department of Defense. Initially, my worries made little difference. I was merely a special assistant to an assistant secretary, and he garnered all the attention. As I moved up the DoD ladder, however, I found myself again coming into contact with many journalists, some of whom I had known for a while, others of whom were new faces.

While the department had its rules about speaking to the press, particularly regarding going "on the record," these were not necessarily prohibitive. One had to contact the Office of the Secretary of Defense's public affairs office to let it know the nature of the request. The public affairs people were exceedingly professional and helpful. One person in particular, Major (now Colonel) Larry Icenogle, often joined my interviews and worked with me on formulating press guidance. He found attending my interviews helped him keep abreast of the latest developments in the Lavi affair; I found his presence useful protection against being misquoted, a common practice among many of the less scrupulous Israeli journalists.

In coming to grips with the matter of press relations, I was influenced by one particular incident, as well as by the behavior of one of the assistant defense secretaries with whom I worked. The incident involved an encounter with an investigative reporter I had known from my CBO days, a brilliant analyst who owned a doctorate from MIT. The reporter had been researching a weapons program that was under consideration in the Defense Systems Acquisition Review Committee (DSARC, pronounced "Deesark"), now called the Defense Acquisition Board. He had contacted me when he learned that I was a member of the committee. Actually I sat on the committee only as Fred Iklé's representative, since he had little time or inclination for the minutiae of defense acquisition matters.

Despite our good personal relations, the telephone call made me nervous; I had never before been asked to reveal any details about my work at DoD. I was still unsure of my position on the committee, both because I was not a member in my own right and because as a special assistant in Iklé's office, I was among the most junior officials at the committee table.

The reporter told me that he knew that some decisions had been made at the most recent committee meeting and asked if I could help him. I refused to comment. He then told me he had a paper outlining the committee's activities, and therefore didn't need my help anyway. He was not pleased with my lack of cooperation. In the event, the paper he had obtained was an early version of a draft decision memorandum, which subsequently was altered. His report was inaccurate and mildly damaging. But I had learned my lesson. I decided to be more forthcoming to journalists, since in that way both our interests could be served. If I steered them away from inaccurate information that they might have received, their credibility would remain intact and the department would not be hurt. If I could feed them background information of which they were not aware and that helped the department, so much the better. These were not particularly novel conclusions—but I believe that every official must reach them on his or her own, before he or she is really comfortable when dealing with the fourth estate.

The second factor affecting my relations with the press, the behavior of that assistant secretary, accelerated my growing recognition of the press's ability not only to break officials but also to make them. This was not, as at CBO, a matter of getting publicity for an institution—the Pentagon had always had more than enough of that—but rather a very personal issue. It is important to recognize that life is hardly a bed of roses for the subcabinet political appointees who daily grease the wheels of government. Burdened with long stressful hours, rewarded with relatively low pay for jobs of significant responsibility, absent from the daily activities—school plays, birthday parties, team games—that make life bearable and family meaningful, the under, assistant, deputy under, and deputy assistant secretaries that populate the capital require other outlets for job satisfaction. For some it is attending parties and traveling—with their wives and children as often as possible. For others it is winning the rat race for perks of other sorts—as exalted as accompanying the president on trips, or as petty as winning special dining privileges. And for those so-called Washington insiders, who may never have left the capital city since graduate school, do not have law degrees, and have worked their way up the political greasy pole, often by serving as staffers on Capitol Hill before moving to the executive branch, a job title which includes the magical word "secretary" often can be a passport to a much-higher-paying post-government job. The problem for such people is that there are several thousand "senior" officials floating around town, all of whom have the same idea. Since the vast majority of these people not only want high-paying jobs but are sufficiently infected by Potomac fever not to want to relocate—ever—there clearly are not enough jobs to satisfy them all.

The key to the successful job hunt is, therefore, publicity. I once worked with an assistant secretary who, despite his rather unpopular political views, established himself as a household name—at least in those Inside the Beltway houses that "count" in these matters—by devoting the better part of most of his working days to currying favor with the press. His daily schedule often read like a never-ending press conference—breakfast with a television reporter, a morning interview with a foreign journalist, a lunchtime address to a group of defense columnists, and so on, concluding with a small dinner he hosted at home with two visiting out-of-town reporters for large papers. He would always pass on to each one a little tidbit—some gossip, a harmless leak, or perhaps, if authorized to do so, a leak of greater consequence. Many of these reporters were as liberal as he was conservative. But he was the lobbyist par excellence—they could not dislike this man even as they reviled his views. And he was eminently quotable. Indeed, the more this fellow was quoted, the more influence he garnered with his own political masters and with politicians overseas, the more his views were sought out, the more he had to offer the press. It was, for him, a wonderful conundrum; he now earns perhaps ten times annually what he did when he first entered the Pentagon. He had but one concern, that the newspapers spell his name correctly; what they said *about* him was of lesser consequence.

These lessons stood me in excellent stead as I parried countless requests for information—not only about the Lavi, but about my private life—from insatiable reporters. I tried my utmost to be helpful to the reporters who contacted me. I had learned long before that, like other people, they appreciated my returning calls quickly—they usually had deadlines to meet, either that day or soon after. Recognizing that there was no way to manage a low profile while the Lavi issue simmered in Israel, I worked with Larry Icenogle to keep our position very much before the public. Since I was in charge of issuing "press guidance" for others regarding the project, I had a freer hand in what I myself wished to say for the record. In any event, support for our work both within the Pentagon and throughout the government was virtually unanimous. Nevertheless, my newfound prominence soon translated into notoriety within the American Jewish community and resulted in increasing anguish, particularly for my family, as the Lavi episode continued to take its course.

I found myself dealing with five distinct clusters of journalists. There were the reporters for the specialist defense and/or high-tech-related press, dailies like *Aerospace Daily* and *Inside the Pentagon*, weeklies like *Aviation Week*, *Defense News*, and *Defense Week*. Then there were the journalists from the mass media, the newspapers, radio, occasionally television. There were also foreign journalists, normally from large-circulation papers overseas. Fourth, there were the Israelis, the print, radio, and television types, all of whom seemed as interested in my personality as in the issues generated by the report. Finally, there were the American Jewish journalists, ranging from top-quality people such as Wolf Blitzer, now with CNN but then with the *Jerusalem Post,* and a host of American Jewish journals such as the New York *Jewish Week,* all the way to scribblers such as one fellow who at the time worked for the *Washington Jewish Week* and who so annoyed me that he became—and still is—the only reporter with whom I absolutely will not speak.

For the Israelis, and to some extent the American Jewish writers, and even, on occasion, the mass-circulation American press, the central question was "How can a nice Jewish boy oppose the State of Israel?" More specifically, once the press learned that I had been ordained as a rabbi (though that was fifteen years before, and I had never used my ordination professionally) it was either "How can a rabbi oppose Israel?" or "Why is a rabbi doing the Pentagon's dirty work?" or both.

I didn't mind dealing with questions of my opposition to Israeli policy, or, for that matter, questions of loyalty. The loyalty issue had never bothered me before, or since, since I have always viewed myself as an American who is Jewish. Yes, I feel a strong sense of devotion to my cultural motherland. Yes, as an Orthodox Jew I long for the Messiah, whose coming will mark my return to the Promised Land. But we Jews believe that the Messiah hasn't shown up yet, and, as a consequence, I do not feel that I must emigrate to Israel tomorrow. As the comic Alan King once put it during some TV show whose name I have long since forgotten—and he wasn't smiling when he said it—"Israel is my mother and America is my wife. You *can* love two women at the same time."

I firmly believe that over the long haul, America's and Israel's interests do coincide. But that is over the long haul. In the short run there can be, and often are, differences between the two, just as there are differences between America and other of her close allies— France, for instance. When those differences arise, I view the issues at hand in American terms. There may be times when I conclude that the administration of the day is harming American interests by disputing the matter with Israel, but my views are still grounded in American interests.

The Lavi issue, in this context, was an easy matter to resolve.

It was not, as its supporters claimed, that America's behavior was harming not only Israeli but also American interests. Rather, the case was quite the reverse. Israeli insistence on building a costly aircraft not only was a waste of American taxpayers' money, but it was against Israel's interests as well. It was unfortunate, but hardly surprising, that the Lavi's supporters did not see things the way I did.

The matter of dual loyalty had another, more troubling side for me, however. That was the Israeli refusal to understand that I viewed matters from an American perspective. Indeed, my outlook seemed so shocking that the mass-circulation *Ma'ariv* ran an article that spring with the headline "Nice Jewish Boy Grounds the Lavi" and a sidebar that described me as a graduate of Yeshiva University (not true), an ordained rabbi, an Orthodox Jew, and a Hebrew speaker. It evidently considered as newsworthy my remarks that "I am an American. Period" as if my behavior were both a revelation to Israelis and an aberration from the American Jewish norm.

The prevailing Israeli view—and that of some Americans, I might add—seems to have been that I was guilty of dual loyalties by putting *America* first. My first loyalty should have been to Israel. I should have viewed disputes between our two countries from an *Israeli* perspective. What was bad for Israel must be *ipso facto* bad for America. Even if it was not, it should be.

That attitude permeated the questions I received from Israeli journalists. In true Israeli style, they were not subtle at all. They hammered away at the issue over and over again, and reported my responses to a putatively horrified Israeli public.

The issue of dual loyalty was further complicated by two other factors—the Jewish-American and Israeli public's clear antipathy to Caspar Weinberger, and the Pollard affair, which involved a Jewish-American naval intelligence analyst who was caught spying for Israel, and which took place at about the same time as we were engulfed in the dispute over the Lavi. These matters deserve a chapter of their own, which follows.

Although I was sometimes troubled by the way I was characterized in the press, I was often more annoyed about the references to my ordination, which had nothing to do with my professional career, than those to dual loyalty, which is an Israeli obsession. I felt that most of the journalists I met were decent people and serious professionals who gave me a pretty fair shake. I had, as I have mentioned, encountered some backroom put-downs and gossip at the outset of our project, but these too had been a minor irritant. What I was totally unprepared for was the onslaught of vituperation that accompanied the report's publication and the eleven-month run-up to our evaluation of Lavi alternatives.

As the battle over the Lavi wore on, I found my skin getting thinner, especially when my family was dragged into the fray, which seemed to be happening with ever-increasing frequency. I started to receive hate mail and nasty phone calls, but got used to them, as I did to being called a "traitor." But it is those incidents that in some way involved my parents, children, wife, and other members of my family that stand out most sharply in my memory, and that point to the depths of uncivil behavior that the Lavi's proponents were prepared to plumb.

Because the future of the plane had become a major issue for the Jewish community, some in that community evidently reasoned that my family was no longer exempt from abuse, and that neither I nor they should be permitted to escape the Lavi even on the Sabbath, even in synagogue.

Some of the annoyances were merely petty. Friday night provided officials from the Israeli embassy who were members of my synagogue an especially opportune time to badger me about the Lavi. Orthodox Jews do not drive on the Sabbath; we walk home from services. My house is located about a quarter of a mile down the road from the synagogue, in a subdivision that is home to most of its members, including its Israeli members. In other words, virtually everyone from the synagogue walked at least some distance along my route home. Moreover, most people mill about in front of the synagogue for a while after the service ends, so as to offer one another the customary Sabbath greeting, *Shabbat shalom*. That meant that those Israelis who did not escort me part of the way home could still waylay me in front of the synagogue while I was exchanging greetings, and that is exactly what happened Friday night after Friday night. The fact that all of these people knew that my family were eager to get home and proceed with the traditional welcoming of the Sabbath—the songs of Sholom Aleichem and the kiddush ceremony—didn't seem to make much difference to them at all. Moreover, as Orthodox Jews, the Israeli embassy people knew quite well that it was improper (many rabbis would say forbidden) to discuss workaday matters on the Sabbath. That didn't stop them either.

One particular embassy official would pick my brains—on the street corner outside the synagogue—about the latest developments regarding our efforts. What were we planning to do next? What was State going to do? Weinberger? Then he would launch into a routine that was common to all of the Lavi's supporters, be they in AIPAC, Israel, or the Jewish community: "Why are you doing this? What's a nice Jewish boy like you involved in this business for? You're not just a government official. You're different, you're Jewish."

I never really felt on the defensive, or in any way pressured. Irritated was a better description. These people simply got on my nerves. Especially on Friday night. Especially this one individual. The last thing I needed, after a long stressful week, was to discuss the Lavi after services. I have been conditioned to keeping the Sabbath for so long that I can literally feel the adrenaline draining from my body as I drive home from work on a Friday evening. All I long to do is have my Sabbath meal and hit the sack. And this fellow, ignoring the cold winter evenings on suburban Maryland street corners that are not noted for their pleasantness, would go on and on about the damn airplane. Still, I could have put up with him had it not been for the fact that these interminable post-services discussions really bothered my eight-year-old son, Reuven, who took matters of religion very seriously. And it was through Reuven, in fact, that this fellow received his comeuppance.

Reuven normally would accompany me home from synagogue, usually holding my hand. During the winter months, he would stand on the street corner in front of the building freezing while the Israeli wouldn't let me walk home. When I finally was released, because the fellow had to get home to *his* wife and kids, Reuvie would bitterly complain to me about him, hurling all sorts of imprecations at the Lavi and its supporters. "I hate the Lavi. Why can't they leave you alone, Dad?" This fellow's badgering—there is a wonderful Yiddish word for it, "noodging," that really describes what he did—persisted on through the spring. Then one Friday evening in late May, as service was ending, Reuvie joyously informed me that I would not be bothered anymore. I asked him why.

"Well, Daddy, you know that so-and-so's daughter is in my class at school." I said I knew that. "So, Daddy, yesterday she was bragging to the class about what a big shot her daddy was, how he is important in the Israel embassy and all that. Well, I raised my hand and said, 'Yeah, but your daddy talks about his *work* to my Daddy on Shabbos, and that's *wrong*. Daddy," he assured me, "he won't bother you anymore."

My son was right. The only words we exchanged after services for the next two years were "*Shabbat Shalom*" and "How is the family?"

That, however, was perhaps one of the few cases where I was able to keep the airplane out of my family's life. I had other encounters in the synagogue that did not end as well. One Sabbath morning, as I walked out of the sanctuary to the cloakroom to get my coat, holding my three-year-old, Saadya, by the hand and followed by my other two boys, I was loudly accosted by a woman whom I had never met before. She said, "If my looks could kill, you would be stretched out on this floor right now." For a moment, I didn't realize that she was speaking to me. She went on, "A person like you should believe in Gehinnom, because that's where you'll go." I still didn't register what she was talking about. I thought I might have stepped on her toe, or jostled her as I left the sanctuary. Then she made noises about "treason," and I caught on quickly enough. She had been speaking in the midst of a crowd milling around the coat rack grabbing for their coats, but by now the place had gone quiet. I must have turned pale. The older boys, who understood what she had said, were stunned. Little Saadya was just frightened. She asked me if I knew who she was, and, trying to retain my wits while Saadya clutched my hand, I told her I did not. She identified herself as the mother of someone who had once been my neighbor in a different part of Silver Spring and who had since moved to Israel, where he had become a lawyer for IAI. The lady then cursed me again, stating in no uncertain terms that I should drop dead.

Eli Avidan, the Israeli political counselor who, to his credit, rarely bothered me about the Lavi, whispered to me, "Ignore her." He ushered me out the door with my shocked children in tow. Curses are not a trivial matter to Orthodox Jews, or, for that matter, to many people. Our Sabbath was effectively ruined; even the traditional shot of whiskey at post-service kiddush could not revive my spirits that day. As for the boys, they simply could not reconcile their school's teachings of love for their fellow men with what they had witnessed that day in shul.

That curse summed up much of what we faced throughout 1986

and the winter of 1987. One Saturday night, my anguished father told me that his brother, his only surviving sibling, had been given a very hard time during his just-ended visit to Israel. My uncle had been told that I was a Nazi, and was foolish enough to tell my Dad. The two of them are Holocaust survivors; they lost their parents and two sisters during the Nazi invasion of Lithuania.

My father, an active member of the American Betar, which staunchly supported the Israeli Likud, was hounded by Moshe Arens's American-based minions, many of whom he had worked closely with for years. On one occasion, he had written a preface to a Likud document that was meant to be circulated nationwide. He was extremely proud of what he had written. As a Betari since his student days, he was delighted to be both active and appreciated for the nearly six decades that he had devoted to his beloved cause. Shortly before the publication deadline, Dad learned that his preface had been dropped. The pamphlet appeared with a preface written by Moshe Arens. My father soon learned that Arens had exploded at the prospect of having anyone named Zakheim associated with a major Likud paper. Needless to say, my father was stung. To his credit, he did not turn on me, which he could have done, since he bitterly disagreed with my attitude toward the Lavi. Never once did Dad utter a word of criticism; to the contrary, he insisted to me that if I felt as strongly as he thought I did, I should follow through on my beliefs. As for Arens, Dad had never thought much of him anyway, and the incident with the preface gave him no reason to change his evaluation of the man.

Of course, the fact that my dad was given a hard time because of me did not prevent the Israelis from giving me a hard time because of my father, often in the presence of other members of my family. I recall one encounter with Meir Rosenne, the rather stiff Israeli ambassador to Washington, at the home of Eli Rubenstein, who was then Rosenne's deputy. In response to my hello, Rosenne, a longtime Likudnik and sometime acquaintance of my father's, barked out within earshot of at least a dozen people, "How could a father like yours have had a son like you?" I must admit, it was hard to think of a quick rejoinder to that one and turn away. My wife, who was standing beside me, guided me to another part of the room, clearly upset by the encounter.

My children didn't escape the hassle either. Chaim, my eldest son, was told by his camp counselor that his father was a traitor. Curiously, Chaim long held the opinion that I was wrong about the

Lavi and should be more lenient with Israel. He often would say as much at dinners we hosted, to the amusement of our other guests. But I was proud that my twelve-year-old had a mind of his own, and I never sought to change his mind. He did that on his own. One day, he came home from school announcing that he was now fully behind me. I asked him what had happened. He answered that the class had launched into a discussion of the program—as part of its focus on current events—and he thought he heard the teacher say that the United States was paying for the Lavi. He wasn't sure that he had heard her correctly, so he asked her explicitly if that was the case. He was told that it was. He then told me that he had supported the plane because he thought the Israelis were spending their own money, and I had no business trying to stop them. "But it's *our* money they're spending," he said. "You're right, Daddy, I support you." Only a parent can appreciate the wonderful feeling I derived from that particular policy shift.

As annoying as these incidents were, I found equally upsetting my abandonment by the Orthodox community of which I had been a part since birth. Time and again I found Orthodox leaders, rabbis, and institutions unwilling to acknowledge that I had any connection with their denomination, with the one major exception of the Satmar Chassidic community, which had no truck with the government of Israel anyway. Again an incident involving a family member stands out among the rest. My brother Josh had inquired of a friend who worked in the public affairs office of a major Orthodox institution with which I had once been closely associated whether it would list my name in one of its publicity vehicles. The reply upset him. "Once the Israelis have nice things to say about him," Josh was told, "of course we'll mention him; until then, we have no connection with him at all." Evidently the USSR wasn't the only place where nonpersons could be reinstated to acceptability in consonance with shifting political winds.

My immediate family also found that our lives had become rather transparent, at least to Israeli readers, a rather unusual development for the family of an unelected official—from another country, no less. When I played in my children's school's father-son basketball game, the Hebrew press even gave details of the number of points I scored (four). And when I attended my youngest son's Purim holiday play, the Israeli press considered this major event on the nursery school calendar to be sufficiently newsworthy for its readers.

One evening, my wife received a call for me from a journalist in Israel who wanted to interview me. My home number was available in the suburban Maryland telephone book, and I often received calls at home. So the call itself was no great surprise. What did surprise her was the journalist's reply when she said that I was out: "Okay, can I interview you?" The journalist then proceeded to ask how she could sleep at night when her husband was quarreling with Israel over such a major issue. "Easy enough," she said. "When you have three active boys, you sleep very well at night." The rest of the interview was conducted along very similar lines, and, sure enough, a few days later, the paper carried the story: "Zakheim's wife has no trouble sleeping despite the Lavi."

9

Two Sides of Dual Loyalty: Cap Weinberger and the Pollard Affair

American Jewry and the Israeli public have long viewed senior administration officials as either "good" for Israel or "bad." Caspar Weinberger was most definitely viewed in the latter category, the bad guy, the black hat—the official who opposed any American initiative that might be of assistance to Israel.

The tradition of black hats and white hats stretched back to the years of Israel's creation. During the Truman administration, the black hats were General George Marshall and James Forrestal—the latter was an avowed anti-Semite—and the white hat was Clark Clifford, who convinced Truman to recognize the fledgling Jewish state. During the Eisenhower years, John Foster Dulles wore the black hat, and there was no obvious counterpart wearing a white one. By the time Jimmy Carter was in office, and the Lavi was becoming slightly more than a gleam in Menachem Begin's eye, the black hat was worn by Zbigniew Brzezinski, the white one by Stuart Eisenstat, Carter's domestic affairs adviser. I happen to be acquainted with both men. Stu Eisenstat, later ambassador to the European Union, certainly has always been a friend of Israel, though I am not aware of his lobbying on behalf of the Israelis from his domestic affairs vantage point during the Carter years. Still, Eisenstat is a discreet man, not prone to showmanship of any kind. He may well have been a strong advocate of pro-Israeli positions in a White House that, despite Camp David, was often less than friendly to the Israelis.

Yet if Eisenstat deserved his white hat status, Brzezinski certainly did not merit being branded as an anti-Israeli desperado. Worse still, he was called an anti-Semite, which he most certainly is not. Zbig Brzezinski was my professor at Columbia. I took his course during my senior year, when I wore my yarmulke at school. Not only was he eminently fair to me, he never made me feel uncomfortable because of my yarmulke, even though I was one of the few students on campus at the time who wore one. My class had only two students who covered their heads at all times. Moreover, I happened to be rather sensitive to any sort of insinuation about my religious practices, particularly as a student in the sixties, when almost anything anyone over thirty said to an undergraduate was viewed as giving some sort of offense. It was wrong for American Jews, and Israelis, to tar Brzezinski with the brush of anti-Semitism; heaven knows there are enough real ones around in Washington who deserve that unpleasant appellation. And it was equally wrong to call Weinberger an anti-Semite.

Cap Weinberger's grandfather was Jewish. Weinberger never hid that fact, even though he himself was raised as an Episcopalian by his Protestant parents. My dear friend the late Rabbi Seymour Siegel, professor of philosophy at the Jewish Theological Seminary, was shocked at Weinberger's treatment at the hands of the American Jewish press and communal leaders. Seymour had known Weinberger when he was secretary of health, education, and welfare, an agency that traditionally has many Jewish bureaucrats and close contacts with many Jewish interests. Seymour simply could not recognize the man as he was now being portrayed in Jewish circles.

Seymour's views notwithstanding, the secretary of defense was the administration's reigning black hat, while Secretary of State Alexander Haig, who had developed close ties to Israel at the same time as Weinberger was making many Saudi friends while at Bechtel Corporation, became the Jewish community's favorite white hat. When Haig left State in June 1982, Shultz took his place in Israel's and American Jewry's favor. Despite his own time at Bechtel, the new secretary of state soon became an outspoken advocate of close ties with Israel when the Syrians torpedoed the 1983 Israeli-Lebanese agreement that he had personally brokered.

Weinberger continued to treat the Israelis with extreme caution.

The caricature of Cap Weinberger as a self-hating Jew, in the tradition of the Weimar foreign minister Walter Rathenau who was renowned for not being able to stomach his Jewishness, was too good for the overwhelmingly liberal and Democratic Jewish community to pass up. Since I was "Weinberger's person," and a Jew myself, the opportunity to get at me, and at Weinberger through me, was equally tempting. Most Israelis and American Jews I met believed that Weinberger was Jewish and was a traitor to the faith. It was therefore not too difficult for some of the more demagogic scribblers in the Israeli and American Jewish press to convince their readerships that I, in turn, was a traitor to my people.

For my part, I was—and remain—proud of my association with Weinberger. He always treated me, and, for that matter, my family on the few occasions he met them, with the utmost decency and courtesy. His decency was virtually self-effacing. I recall once getting a call put through from my secretary: "It's Weinberger," she said. I wasn't used to hearing directly from the SecDef, as we called him, so I asked "Which Weinberger?"—I knew a few Jews of that name. "Weinberger," came the reply. I recall being annoyed to have been interrupted in the middle of something that must have been important at the time, but not important enough for me to remember what it was. "Yeah," I said into the receiver. "Oh, Dov, it's Cap Weinberger here," said the voice at the other end, as smoothly as if he spoke to me every hour. And he went on to tell me what he wanted. The incident lingered with me, not with him, as it would have with someone petty or anti-Semitic. But Weinberger is neither petty nor anti-Semitic.

Still, just as Weinberger has gone down in Jewish lore as a "hater of Israel," so has my association with him never failed to evoke some comment from any group of Jews I meet, particularly when on a speaking engagement. "Do you still see Weinberger? Wasn't he anti-Semitic? And what about his treatment of Pollard?"

Which brings me to the second factor that sucked me into the dual-loyalty debate: the Pollard affair.

Jonathan Pollard was a low-level Jewish-American intelligence research specialist who was arrested by the FBI in November 1985 on charges of spying for Israel, apparently for rather large sums of money. The circumstances of his arrest were quite melodramatic. He and his wife sought, and failed to obtain, asylum at the Israeli embassy in

Washington. The Israelis vigorously denied that he had acted for them, terming his efforts part of a "rogue operation." Pollard was the proverbial spy left out in the cold.

I never met Jonathan Pollard. My reaction to his behavior was, at the time, unalterably negative: I don't like spies of any stripe, whatever their race or creed. His arrest briefly attracted my attention, but I was too busy with other matters. On June 5, 1986, however, Pollard pleaded guilty and was subsequently sentenced to life in prison.

I have often been asked whether the Pollard business affected my status at the Pentagon because I was Jewish. For although I was not the most senior Jew in DoD—that honor went to Assistant Secretary Richard Perle—I was certainly one of the more visible Jewish officials at the department. In fact, it was not with my non-Jewish colleagues that I had problems, but, as in so many other aspects of the Lavi affair, with my Jewish acquaintances outside DoD, as well as with the Jewish community at large and, of course, the Israelis.

Virtually from the day he was sentenced, Pollard sought to portray himself as a Jewish martyr whose efforts on behalf of Israel deserved approbation, not renunciation. Pollard quickly became something of a hero to the rank-and-file American Jewish community, though the communal leadership was very slow to speak out in his favor, or, for that matter, to speak out about him at all. But to the Jew in the street, bombarded by a very clever public relations campaign organized by Pollard's family and close supporters, Pollard was being "set up" by that anti-Semite Weinberger. While the Israeli government was embarrassed by Pollard and continued to disown him, Pollard increasingly became something of a hero to the Israeli public. And, in the course of doing so, he became my antithesis. As Israei television put it in an interview shortly after I left the department in 1987: "Jonathan Pollard is seen in Israel as the American Jew who helped Israel; Dov Zakheim is seen as the American Jew who hurt Israel."

Much continues to be made, especially within the Jewish community, of Pollard's motives and the purported harshness of his treatment since his imprisonment. I have never involved myself in these matters, although I was on occasion approached by various people, including Israelis, to lend my voice to the chorus calling for leniency. I have no real knowledge of the details of his case since his arrest and do not feel qualified to speak on the matter, other than to reiterate that my visceral repulsion at his behavior remains unchanged.

Evidently the position of the State, Justice, and Defense departments as well as of the Clinton White House regarding Pollard's release has also not changed: in the spring of 1994, President Clinton, acting on the recommendation of his law enforcement, national security, and intelligence advisers, refused to pardon Pollard, whose hope for early release from imprisonment rests with the parole system. Ironically, Cap Weinberger, who had been tarred as the evil demon of the Pollard case, was quoted as being in favor of leniency. Those still serving in government thought otherwise.

When Israeli television contrasted me unfavorably with Pollard during a 1987 telecast, it was done as a preface to the question "How do you feel about such a characterization of your activities?" My answer then is the same today. Pollard was a spy—whatever is said and done about his sentence or his treatment cannot alter that fact. If he was a true Zionist, he had the option of moving to Israel as so many other American Jews have done, who also have never hurt, and often assisted, America, the country of their birth, both before and since their *aliyah*. He chose not to go on *aliyah*. Instead he chose to betray his country. That there were not further anti-Semitic repercussions, as many American Jews feared, was due to the maturity and essential decency of all Americans, who could tell a rotten apple when they saw one.

10

Run-up to Another Study Visit

Once Defense Minister Yitzchak Rabin had agreed in May 1986 that we undertake an alternatives study, the issue arose as to how not only to produce the report, but also to ensure that it was the knockout blow against the airplane. I felt that we had to present the Israelis with the widest possible set of alternatives. We therefore had to offer each of the major aircraft manufacturers, and not only General Dynamics, whose F-16 was the Lavi's most obvious competitor, the opportunity to be included in the alternatives study. I also felt that we needed to impose a new, tougher go-slow on all Lavi contracts, so as to increase the pressure on the Israelis and demonstrate that the United States no longer would fall over itself to satisfy the program's needs.

We first had to reconstitute the Lavi team. The fact that the initial study had been an interagency effort clearly had impressed the Israelis; we needed the same degree of government-wide support if we were to convince the Israelis of the viability of an alternative to the Lavi. Accordingly, with Fred Iklé's approval, I established a new "Lavi Alternatives Steering Group," with roughly the same membership as the original Lavi team.

I encountered no significant opposition to my strategy, though some elements on the Air Staff felt that another study effort would seriously drain the air force's analytical resources. The original air force members of the team, as well as the other agency representatives, were eager to get started and shared my view about the importance of the study. As for the go-slow on the contracts, that idea had already been foreshadowed in DSAA the previous month.[1] Most important, Cap Weinberger approved the plan. I briefed him on Monday, May 12, outlining our approach to the alternatives and our go-slow strategy on contracts. He quizzed me as to which plane was really more effective, the F-16 or the Lavi. I told him I was convinced that the F-16 was the better plane. "Then maybe we should let them build the Lavi after all," he deadpanned, and for a moment I thought he was serious.

Our slowdown of Lavi-related contract approvals soon found its way into the Israeli press. Even Cap Weinberger was quoted as stating that Israel would do better to terminate the project and face a public outcry rather than to continue to waste its money.[2] Everyone had long known that Weinberger felt that way. What was significant was that he was now openly voicing this view. It signaled a new hardball approach to the project that left no doubt as to the readiness of the U.S. government, or at least the Department of Defense, to push for the program's termination.

Weinberger was not being mean-spirited about the program, as many of his detractors sought to portray him. The Israelis had made a habit of pushing for spending approvals on elements of the program well ahead of schedule, so as to lock in spending on the Lavi for future budget years. It was to this practice that we most objected. We made it quite clear to the Israelis, both publicly and privately, that such an approach was no longer acceptable. Given both cost and schedule uncertainties surrounding the program, it made little sense to approve contractual arrangements well in advance of their being implemented. If the Israelis wanted us to approve new Lavi production contracts, they needed first to recalculate the plane's costs in a realistic manner.

As was their wont, the Israelis expressed bitter outrage, and sought through their various official and unofficial channels in Washington to have the policy reversed. Abraham Ben-Joseph of the New York purchasing office protested the policy to me when we met on May 20. The following day I met with the MoD's senior economic adviser, Zvi Tropp, who was also unhappy with our policy. Tropp was heading a small delegation to Washington that was attempting to convince anyone who might listen of the correctness of their cost

estimates and the flaws in ours. They hoped thereby to free the contractual logjam. They got nowhere; we insisted that in the absence of
a clear long-term cost and program plan for the system, accelerated
components acquisition made little sense.

Although they failed to convince us, the Israelis were determined
to free the contracts and to vindicate the program. The pressure was
so great that for the first time since I had begun working on the Lavi
issue, I fully expected to be pulled off the case. Weinberger had seen
the press account of his stated opposition to the program, and the
report had carried several of my own statements regarding the plane's
dim prospects. With Israeli protests mounting to a crescendo, if ever
there was a time to shoot the point man, this was it. But no one in
the Pentagon was particularly interested in giving way to the Israelis.
Quite the contrary; Cap Weinberger and Assistant Secretary Rich
Armitage in particular were very sensitive to any attempts to undermine me personally. Instead of being made the scapegoat as a result of
Israeli pressure, I constantly received support and encouragement,
especially from Rich.

During the week that I met with Tropp, I also got together with
Colonel Mike Foley of the air force to plan our next visit to Israel
and our overall analytic approach to the alternatives effort. We preferred not to issue another report. Rather, we would produce a briefing that could be presented to Rabin, his ministerial colleagues, and
the senior MoD people.

Our effort borrowed heavily from my experience at the Congressional Budget Office, which among its other analytical strengths has
developed a well-deserved reputation as a source of unbiased force
alternatives and their budgetary implications. I turned for help to Bob
Levine, who had been deputy director of CBO when I first came on
board there in 1975. For additional advice and suggestions, I looked to
two of the nation's top economists, Stan Fischer of MIT and Herb
Stein, former chairman of the Council of Economic Advisors, now at
the American Enterprise Institute (AEI), a Washington think tank.
Both men had played a leading role in the American effort to assist
Israel out of its inflationary mess in the mid-1980s, and both were
keenly interested in the Lavi debate. Of the quartet, however, (the
fourth person serves in the Clinton cabinet) only Bob Levine was
prepared to allow his name to be mentioned in the context of what
I was trying to do. The others felt that it would be impolitic for them
to do so. What I needed was advice, assistance, and, in some in-

stances, comments on our draft report, and it was easy to accede to their wishes. We never mentioned their involvement—not even Bob Levine's.

Together with my deputy, Dick Smull, Mike Foley and I outlined a detailed schedule for the report. We anticipated that the "terms of reference" would be approved by June 30, that the preliminary analysis would be completed six weeks later, and that the draft would be completed by September 3. As we had done in the case of the first study, we gave other interested parties, both inside and outside DoD, a chance to "coordinate" on the study. We allowed ten days for the study to make the rounds of the government; we indicated that if we had not received comments from recipients by the due date, we would assume that they concurred with the report.

The schedule then called for a final round of analysis and a final draft to be completed on September 22. We allowed three days for final coordination, to give agencies a last chance to speak up or forever hold their peace; and we anticipated handing the complete version to Caspar Weinberger on September 29. We hoped to brief the Israelis before October 15, or just after the Lavi was meant to make its first flight. As things turned out, the first flight was delayed by a few months, and we did not get to present our report until January 1987.

We expected that the basic work on the study would be undertaken by industry, with the air force Systems Command's Aeronautical Systems Division (ASD) at Wright-Patterson Air Force Base, near Dayton, Ohio, providing quality control. We also assumed that the navy would help with whatever material was required to evaluate the F-18. In fact, we later received a copy of a note from Air Force Secretary Pete Aldridge to Weinberger indicating that the air force planned to approach the navy to coordinate their efforts.

We determined that the study's alternatives should all include an American airframe and Israeli avionics. We did not want to peddle an off-the-shelf American fighter, since the Israelis knew all about that option anyway. We reckoned that the alternatives should at least include the F-16, F-18, and the F-20; we later added the AV-8B and the F-15.

As for the content of Israeli avionics and electronics, we felt that the package should go further than that which General Dynamics had outlined to Defense Minister Ariel Sharon. At a minimum, we hoped the alternatives would incorporate a radar, weapons control systems, electronic countermeasures (to spoof enemy systems, for example, radars, so as to provide friendly aircraft with electronically generated protection), electronic counter-countermeasures (to overcome any attempts by the enemy to spoof friendly aircraft), cockpit displays, information processors, and, if relevant, the Pratt & Whitney PW1120 engine.

We also imposed a cost ceiling to our alternatives. To make them comparable to the Lavi program as it had been redefined by Yitzchak Rabin, we felt that the alternatives should not exceed $500 million (1984) dollars annually. In so doing, we postulated the "broad" interpretation of Rabin's cap by allowing for inflation, but we did not go so far as to assume that the *average* cost would be $500 million, which would allow potentially massive peaks and valleys in the program's funding profile.

One key question was whether the cost estimates that industry would provide for the report would have any relevance to those that might actually be quoted to an Israeli purchaser. We asked for, and received, an undertaking by the contractors that the estimates that they provided to us would be within 10 percent of what they would demand from the Israelis. Of course, those estimates postulated very specific packages. Any change, which was exceedingly likely, would afford the contractors leeway to increase their asking prices by a significant margin. We knew that the numbers were not as tight as we would like, but the "10 percent promise" was the most that we could hope to receive from industry. To offer a guaranteed price for an undefined product made no business sense, and would not have been credible in Israeli, or American, eyes anyway.

We attached one other major criterion to the alternatives: we wished to minimize their negative impact on the Israeli workforce currently employed on Lavi or Lavi-related programs. One of the most politically compelling arguments that had been made in favor of the Lavi was that its cancellation would put thousands of highly skilled workers on the breadlines. Israel had yet to recover fully from its economic crisis. Its hyperinflated economy was devastating for workers on fixed wages. The prospect of unemployment was daunting; it would likely lead to increased emigration, which was Israel's traditional way of coping with its economic difficulties. The governing philosophy in the Histadrut, the powerful national labor union, was that it was better to continue with the expensive Lavi than to have no jobs at all.

The key to our effort clearly was the group of U.S. defense contractors, but we found them reticent about cooperating with us on the project. Northrop, which had so bitterly opposed the Lavi in the early 1980s, actually contemplated withdrawing its analytical support for the study, but eventually decided to participate. The company was in the midst of its B-2 project, as well as others, and had not been able to sell its F-20, which was an outgrowth of the highly successful F-5. Northrop clearly hoped the Israelis might choose the F-20 as a Lavi alternative, but the history of its opposition to the Lavi did not bode well for its prospects. To the extent that it would profit from a sale to Israel at all, Northrop was likely to do so via the F-18, in which it had a 40 percent share.

McDonnell Douglas was the prime contractor for the F-18, just as it was for the F-15 and the AV-8B Harrier, all candidates as Lavi replacements. McDonnell, for its part, was only slightly more prepared to cooperate actively than Northrop was; initially it would do so only through the Aeronautical Systems Division. Any figures or estimates were thus the responsibility of ASD and not of McDonnell.

McDonnell Douglas had as much good reason to be camera-shy as Northrop. The Israelis were courting the contractor with great assiduousness. Officially, they said they were seeking help with the Lavi's avionics integration, with which IAI was experiencing difficulties that it desperately hoped to overcome.[3] There was a more subtle purpose to IAI's attentions. The company, and more important, the MoD, hoped that with an American partner, prospects for sales both to the United States and to third parties would brighten considerably. Israel desperately needed a big production run; its plan for producing three hundred Lavis was clearly in jeopardy. Only with an American partner could Israel hope not only to "market" the air force but also to obtain the necessary permission to transfer American-built technology to third parties.

IAI put its top guns to the job of wooing McDonnell Douglas. Early in May 1986, Lavi project director Menachem Eini and Moshe Keret, the president of IAI, traveled to McDonnell headquarters in St. Louis to discuss a coproduction scheme.[4] The American company was noncommittal. It wanted no part of the Lavi, but it was not about to alienate IAI.

Similar conditions, considerations, and arrangements prevailed with respect to our interaction with General Dynamics, which, as

manufacturer of the Lavi's leading competitor, the F-16, was especially sensitive about appearing to be in league with the Pentagon.
Vern Lee, a soft-spoken but hard-nosed Texan who had worked for
years on the Israeli F-16 program and was now F-16 program director
for Israel, made it clear to me in a formal note dated April 14, 1985,
that he did not want to be enmeshed in the Lavi dispute. Lee followed the project closely, receiving press translations and on-the-
spot analysis from GD's Tel Aviv office. Like McDonnell Douglas,
which had sold F-15s to Israel, Lee recognized that cooperating with
us could leave some hard feelings with the Israeli air force. He anticipated future sales to Israel whether or not Israel procured the Lavi,
and he was not about to let the Lavi disrupt his sales prospects any
more than it already had.

Together with McDonnell Douglas, GD knew that many Israeli
air force officers were not interested in the Lavi. Moreover, officials in
both companies recognized that were Israel to require another plane,
MoD director general David Ivri would have no choice but to swallow
his hard feelings over the Lavi and turn to the United States, and
therefore to either of the three companies that were offering alternatives to his preferred program. By not obstructing our efforts yet
at the same time not actively assisting them, the two major contractors could, and did, manage to avoid alienating either side in the
controversy.

Grumman, producer of the A-10, was in a rather different position. It stood to lose the most from a cancellation of the Lavi project
and had no incentive whatsoever to support our work. On the other
hand, Grumman very much wanted to know what we were up to.
Therefore, in return for openness on our part, the Grumman senior
staff made every effort to brief us on their own cost estimates—which
tended to support the Israeli estimates far less dramatically than the
Grumman people themselves realized. I had no problem with such an
arrangement. We had nothing to hide; on the contrary, we wanted as
much visibility on all sides of the issue as the various program proponents were prepared to offer.

The Grumman estimates had actually been completed the previous December. The company's objective was to deal with the controversy over the cost estimates by generating a set of its own cost
models to obtain a unit flyaway cost. Grumman prepared an extensive briefing that outlined the history of modeling techniques for
obtaining the cost estimates of new aircraft programs. The company

then described its own model, called TOPAS, for Total Parametric Aircraft Costing System. The model took into account a variety of inputs relating to the design of the aircraft and its specified performance (such as speed and weapons load). It also accounted for what it termed "programmatic" elements. These related to the initial dollar costs, the year the technology was initiated, the year of first flight, the production years, the number of prototypes, engine cost, a choice from among three cost estimates for avionics and armament, and a choice from among production rates of three, six, and twelve aircraft a month. Since the Lavi was actually to be produced at the rate of twenty-four annually, none of these rates reflected the higher cost of production at such a low run. Nevertheless, using a variation of its postulated avionics package, Grumman found itself midway between our estimate and that of the Israelis, aggregating the unit cost to $18.1 million. This estimate used Grumman's labor rates, which the company showed were anywhere from 50 percent to 100 percent more than those reported by the Israelis.

Meanwhile the Lavi was moving ahead, even with our contractual slowdown. Early in May, IAI announced that its Lavi prototype had completed the first maximum-thrust run of an installed engine, about one month ahead of schedule. The company advertised that the plane would be "rolled out," that is, made public, on July 21. The first flight was to take place before the end of September. The plane's supporters hoped that by maintaining an accelerated schedule, they could give the lie to our concerns about "schedule risk," and at the same time fan popular nationalistic support for a plane that would no longer be a paper design but something that the public could actually see flying.

Israel also counterattacked on the issue of the contracts. As was so often the case, the Israelis threatened to force the administration's hand via Capitol Hill. When Weinberger had held up contracts relating to composite materials for the Lavi several years earlier, the Israelis had asked a number of congressman and senators to write to the secretary. In part because of pressure from State, in part because he did not want to confront the Congress, Weinberger had given in, and no letter was actually sent.[5] The Israelis left no doubts that they were ready to march up the Hill again.

As the rollout approached, it became clear IAI intended it to be a major "event." The Israeli press reported that Vice President George Bush might attend; over one thousand people were expected to join him. Although I was ribbed in the Pentagon about getting a front-row seat at the rollout, I didn't exactly expect an invitation to IAI's main event.

I was soon on my way back to Israel, however. Throughout the late winter and early spring we had progressed on the navy study, and we needed to collect additional data in Israel. We had already received briefings from Rockwell International, which hoped to develop the Dolphin submarine's combat system, as well as from Litton's Ingalls Shipyard in Mississippi, Bath Iron Works of Maine, and California's Todd Shipyard, all of which were competing for the contract to construct the Saar missile boats. Bath appeared to be the least interested of the three. Litton offered the Israelis not only construction of the hull, but development of its combat system as well. Todd desperately wanted the contract; the yard was teetering on the verge of bankruptcy.

 The Israelis, as is their wont, were very cagey about their preferences. They egged on all the potential bidders, not only the Americans, but also the German yard that was competing for the submarine contract, HDW of Kiel. Later, when Thyssen of Emden joined the bidding in June, the Israelis simply pushed at it as well. For our part, we pressed on with our estimates, trying not to bias them in favor of the numbers put forward by one or another contractor.

Our trip to Israel, though it really was meant to be a fact-finding mission for the naval study, also seemed to offer an excellent opportunity to take Israel's political pulse with regard to the Lavi. The Israelis were less anxious to hear about our plans for comparing the various aircraft than to achieve some sort of compromise on our varying cost estimates. The Lavi team seem to have expected us to bring a set of revised estimates in response to their latest critique, but we felt that we really had very little to revise. We did still hope for more information on the Lavi, but we expected as little from the Israelis as we knew they would receive from us. When it came to cost estimates, both sides were at an impasse that was unlikely to be bridged.

That was not good news for the Israelis. They were beginning to feel real heat from their own press and public and needed to reassert their credibility, which seemed to be dissipating with every passing day. As the *Jerusalem Post* noted on the day of our arrival in Israel, despite the preparations for the rollout, "below the surface the mood is less than buoyant. The constant attacks on the project and the question marks over the Lavi's future have affected morale and made it all but impossible for the IAI to enter into serious negotiations for the project with an American partner."[6]

One sign of the changing nature of the domestic debate was the outburst of vocal opposition to the project on the part of Yossi Sarid, a member of the Knesset from the left-wing Ratz Party. Sarid, who served on the Knesset's foreign affairs and defense committee, submitted an "urgent" motion on May 22, demanding a debate on the need to call off the Lavi project because of the lack of resources in the defense establishment. Sarid was no great hero to the military. His intervention resembled those of American legislators known for their opposition to defense spending who would seize upon the supposed interests of the military to justify calling for the cancellation of an expensive weapons system.

Sarid and his party were slowly moving away from nuisance status to political respectability (although they did not really "make it" until they joined Yitzchak Rabin's governing coalition, and Sarid became a minister, in the summer of 1992). His opposition was therefore more than merely that of an antidefense gadfly. It signaled that the Lavi was now a sufficiently well understood issue in the public mind that politicians could score points from their opposition to it. It also meant that the Labor Party would have an increasingly more difficult time pacifying its left wing, which shared Sarid's views on this matter and several others besides.

The Israelis were soon to feel the heat in other ways as well. Prior to our departure, I met with officials from State's Near East and Politico-Military bureaus to ensure that the two agencies were marching in lock step on the alternatives issue. We also "coordinated" on letters that both Weinberger and Secretary of State George Shultz were to send to Defense Minister Rabin and Prime Minister Peres, respectively, outlining the need for a more forthcoming Israeli attitude to our efforts. The Shultz letter was a first; the secretary had not yet taken a public position on the airplane, other than to support it.

Although I had always benefited from excellent cooperation with the State Department at the "working level," that is, from officials at or below the rank of deputy assistant secretary, we had long failed to obtain any public statement of support from George Shultz. Shultz had not yet felt ready to take on the Lavi. He was in the midst of overseeing a radical overhaul—and bailout—of the Israeli economy and did not wish to complicate matters with a spat that was very much the domain of the Defense Department. Indeed, as he later wrote in his memoirs, Shultz viewed the Lavi as part of his overall effort to revive and stabilize Israel's economy, which was not a particularly surprising perspective for a professional economist. His view merged with that of Arens, at least superficially, since both men stressed the importance of preserving the country's high-technology workforce.[7]

On the other hand, reports of Israeli stubbornness on the issue could not have failed to reach the secretary of state. I suspect that he intended his letter, which yet another *Jerusalem Post* story rightly described as "friendly but firm," to be more a plea to the Israelis somehow to reconcile their differences with us than a threat to terminate the project.[8] Nevertheless, the fact that Shultz was prepared to intervene at all on the matter was certain to chill the hearts of the Lavi project team.

The news was not all bad for the Lavi's supporters, however. They still had the prime minister and the cabinet firmly in their camp. Shimon Peres assured the cabinet at its regular Sunday-morning meeting on May 26 that he remained committed to the Lavi. The cabinet reaffirmed its support of the program, though not without the misgivings of several ministers. Among them, we were sure, were Finance Minister Yitzchak Modai and former defense minister Ezer Weitzmann. This show of support was to be repeated, like some political mantra, over and over again during the following twelve months.

With the navy analysts already in Israel, I did not take a large group with me on the plane that the air force furnished to us. It did include Colonel Mike Foley, who was leading the Lavi alternatives analytical effort for the air force, and Captain Linda Hardy, once again on the point for the detailed work. I must also note the invaluable support that the air force gave us in ferrying us from Europe to Israel on its jets. It was not simply a matter of avoiding the usual hustle and bustle that is Ben Gurion Airport. Even with the streamlined arrival procedures that the Israelis offered us in conjunction with the air force flight, we were still besieged by reporters, who often could ask the most inane questions. Our arrival would have been infinitely more uncomfortable had we flown commercially and been at the mercy of the press from the moment we landed until we finally left the airport. In addition, we sometimes went directly to meetings upon our arrival, and the air force flight ensured not only that we were fresh for those encounters, but that we had been able to conduct uninterrupted strategy discussions throughout the flight.

We arrived in Israel on Friday, May 30, 1986, and once again the weekend was my own. Soon after my arrival, I received a telephone message from an old classmate of mine who had emigrated to Israel some years earlier. I was delighted to hear from him and quickly returned the call. I hadn't spoken to him since we had graduated from high school, and I was interested to learn about his life since then. He was also curious, and threw all sorts of questions at me, about my work, what I did, and how I got to be doing what I was doing. I answered in a general sort of way, and then asked him a question or two. Suddenly, the conversation came to a screeching halt. All I had done was ask him what he did. "I'm an engineer," he had answered. "Great," I said, "so who are you with?" Silence at the other end. I thought he hadn't heard me, so I put the question a second time. He had heard me well enough. "Oh, an Israeli company." I had figured as much, so I asked which one. Silence again. By now I had guessed which company. My erstwhile classmate squirmed for a long time before admitting that he worked for Israel Aircraft Industries. He really didn't want to talk at all about himself, his family, or anything else, and evidently realized that it no longer would be fruitful to ply me with questions about the Pentagon or the Lavi. I never heard from him again.

Our arrival in Israel was greeted by a new spate of editorial comment and public debate about the Lavi program. Naval modernization, the reason for our trip, attracted only a modicum of commentary. The cabinet's May 26 decision had done little to dampen the debate. The mass-circulation *Ma'ariv* encapsulated growing public skepticism by noting that the burden of proof was on the government to demonstrate that its figures were correct and that ours were wrong. *Ma'ariv* added that there was opposition to the project from within the prime minister's office, which indicated to me that Peres's economic adviser, Amnon Neubach, felt sufficiently confident about the tide turning against the plane that he could give more explicit, and critical, backgrounders to the press.

Another cause for press comment prior to our arrival was the release of a major review of the Lavi issue by a senior researcher at Tel Aviv University's Jaffee Center for Strategic Studies. The report forecast that as a result of the Gramm-Rudman-Hollings legislation,

Israel might have to live with reductions of $300 million in security assistance. In the "worst case," the study added, those reductions could be twice as large. It concluded that Israel simply could not afford the airplane.

The report raised eyebrows for several reasons. The Jaffee Center is Israel's first, and at the time arguably its foremost, strategic studies think tank. It had an international reputation for balanced and forward-looking analysis on major Israeli security issues; for example, it was among the first institutions to issue serious proposals for a potential settlement with the Palestinians. It was led with a firm hand by a former chief of intelligence who was known to be close to Rabin. Given the center's status, it was hardly surprising that the press played up the report, which was likely to have a considerable impact on mainstream Labor Party thinking.

Rabin angrily rejected the Jaffee Center's report and conclusions. Nevertheless, since he had only recently imposed a ceiling on the project's expenditures, his outrage may have been more for the benefit of his party's Lavi supporters than from genuine irritation at the study's findings.

Among the more interesting published reactions to the Jaffee Center's report was an op-ed piece in the *Jerusalem Post* on the day of our arrival. In a scathing analysis, the author cited the study to support his assertion that it would be exceedingly difficult for the United States to increase its security assistance budget if our estimates proved correct. He went on:

> It may well be that IAI's cost estimates are a better forecast than those of the team headed by Zakheim. But the point is that successive Israeli governments and five defense ministers who had to make decisions on this project have never had before them an independent cost evaluation *before* [his emphasis] each of the crucial decisions that created successive *faits accomplis.* . . . Even on IAI's cost estimates, the project is a millstone around the economy's neck.[9]

The op-ed was but one of three that filled an entire page of the *Post*'s Focus section in the edition that appeared on the weekend of our arrival and anticipated our upcoming talks. Like all Israeli papers, the *Post* gets its largest readership on the weekend, which in Israel takes place on Friday and Saturday. With no papers published on Saturday in Israel, it is the Friday edition that is the equivalent to the American Sunday paper.

The Focus section included the above-noted piece outlining the despondency at IAI, as well as a report from Wolf Blitzer, writing from Washington, that emphasized America's concerns for other Israeli programs that were vulnerable to the Lavi's increasing appetite for American security assistance dollars. Blitzer discussed our slow-down of Lavi contracts, which he estimated totaled as much as $60 million. He cited Pentagon "officials" who said that we wanted to see whether Israel really was going to place a "cap" on Lavi-related spending. The link between the slowdown and the Israeli cap on spending was only a cover story; as irritating as it was, it enabled us to avoid causing even more resentment by stating the obvious, that we wanted to maintain pressure on the program. We had not the slightest doubt that Rabin would abide by his cap; he had given us his word. What we did not know, but hoped to learn in Israel, was just what the cap meant. As was so often the case in our search for information, we were to be disappointed once more.

Despite the press uproar that accompanied our arrival, our actual meetings began on a far more congenial note than in the past, because they primarily involved the navy and not the air force. We knew that the Haifa shipyard was lobbying for a piece of the new program. The shipyard preferred to build the patrol boats, but we wanted those to be built in the United States. Our shipbuilding industry was in serious trouble despite large increases in the U.S Navy's shipbuilding program, which had resulted from Navy Secretary John Lehman's unyielding pressure. The Todd yard in San Pedro, California, was particularly vulnerable. Many in and out of the navy and the industry reckoned that unless Todd won the Israeli contract it would have to go out of business. Bath Iron Works, which had a lucrative destroyer contract, as well as the Ingalls shipyard in Mississippi, which was slated to construct both destroyers and large U.S. Marine amphibious helicopter carriers, also remained very interested in the project.

In addition to our wanting to build the corvettes in the States, it seemed a cleaner solution to have Israel build the submarines. To do so would avoid having to contend with John Lehman and his blue-suit supporters in the navy who were violently opposed to having diesel submarines constructed in the United States. Not surprisingly, John had pushed for this solution to Israel's submarine dilemma.

The Israelis clearly could not build the first sub: they needed to be trained by the Germans. We therefore examined the option of having the Haifa shipyard build the second and third subs after a German yard had built the lead ship. The Germans would continue to remain involved in the program, providing technical support for the construction of the remaining boats. In addition, the United States would build the submarine combat system. Both Rockwell International and Litton, the owner of the Ingalls shipyard in Pascagoula, Mississippi, indicated strong interest in the project. The total cost appeared likely to exceed $1 billion: our team and the Israelis were within 10 percent of each other. Several issues remained unresolved, however. First of all, it was not clear to us that the Haifa shipyard could build the submarines efficiently. It was unlikely that Israel would find a foreign market for its submarines. Was it worthwhile spending tens, perhaps hundreds, of millions of dollars to convert the Haifa yard for the manufacture of but two submarines?

But we faced a second problem as well. Even if all three subs were to be built in Germany, how would they be paid for? It was one thing for the United States to permit its security assistance dollars for Israel to be spent in Israel. It was something else for it to permit those dollars to be spent in Germany. There was some precedent for the United States' spending security assistance dollars in a third country. Washington had also funded El Salvador's procurement of some equipment in Europe that was unavailable in the United States. But those programs involved relatively small sums. Israel's submarine project was likely to exceed $500 million.

Together with the Israeli navy, we wrestled with the issue throughout our stay. Everyone was in good humor, however, in contrast to our tense meetings with the IAI people and the Lavi team. The Israeli navy was a really laid-back bunch. The smallest of Israel's services, the navy viewed itself as an elite group that was underrated by the public at large but made a major contribution to Israel's security. In fact, the navy's stellar record in the 1973 War, when its gunboats defeated the Egyptian and Syrian navies while sustaining no losses, became a textbook case for students of naval warfare everywhere.

On the Sunday morning after our arrival, the navy hosted us in Haifa and, to make a point about its submarines, took us on a "patrol." I had never been underwater in anything so small, except for the mock-submarine ride at Disney World. Whereas our smaller submarines displaced several thousand tons, the Gal class displaced only a few hundred. One felt every move the sub made; it was like flying on a small airplane, except that we were underwater. The navy offered us no sick bags.

All the men we met stressed the dangers of structural problems. It was not hard to be impressed by the risk the submariners ran in boats so small. Perhaps they were simply conducting a PR exercise—the size of the subs did not necessarily mean that there were structural problems or that those problems could not be dealt with—but it was effective PR. We were also given a tour of the Haifa shipyard and shown the Israeli patrol boats, including the Saar IVs, which were smaller than the SAAR Vs were meant to be.

I enjoyed interacting with the navy's officer corps. Avraham Ben Shoshan, the navy commander, was a tall, voluble Bulgarian Jew with a Biblical beard. He could easily have served as Charlton Heston's understudy in *The Ten Commandments*. His deputy was a good friend of mine, Micha Ram, formerly the Israeli naval attaché in Washington and heir apparent to Ben Shoshan. Both Micha and Avraham were possessed of a wry sense of humor, which manifested itself in a variety of ways. Once they got the attention of our entire team by astute arrangements that ensured that my most junior assistant, Mike Bolles, would always be seated next to Micha's aide, Lieutenant Limor. Limor was stunning: she had dark hair, dark eyes, high cheekbones, an upturned nose, and a fashion model's figure. And she knew how to flirt. Poor Mike tried so hard to keep his eyes off her and to follow our discussions. But it was a wrenching effort for him, and we and the senior Israelis positively delighted in his predicament, which, I must say, he took in the best of spirits.

We could not resolve the financing issue while we were in Israel. We all agreed that the most practical approach was to get everyone to consent to the content and nature of the program and only then to address any remaining questions relating to its financing. We returned to Tel Aviv that afternoon for some meetings with the Lavi team in anticipation of our discussions with Rabin the next day.

On Monday morning, therefore, we were back meeting with Rabin. Since our visit had been driven primarily by the naval modernization effort, I did not attach any particular importance to the meeting beyond the special significance that resided in any encounter I had with the Israeli defense minister. That, however, was not the view of the Lavi team. They convinced themselves that the meeting was "crucial" to the program, and apparently hoped against hope that in light of the written critique that they had sent us, I would indicate to Rabin that we had moved closer toward their numbers.[10]

The meeting was certainly not one of my more pleasant encounters with him. Whether it was for his staff's consumption, or quite genuine, or simply a reflection of other things on his mind, Rabin came across as especially dour and unfriendly. Ambassador Tom Pickering handed Rabin a letter from Cap Weinberger, which restated our concerns about the Lavi program. Rabin raised the matter of the blocked Lavi contracts, saying that he, for his part, had agreed to the alternatives study. Pickering and I agreed to try to unblock them, but I was very cautious in making a specific commitment. I was not at all certain that I could change Weinberger's feelings on the issue.

During our meetings at MoD we reviewed my team's differences with Rabin's staff over the cost estimates. These discussions were also rather unpleasant. The Lavi people seemed somewhat self-satisfied about the GAO study, which they fully expected to vindicate their estimates. They shed no light whatsoever on Rabin's cost cap, regarding which the minister himself had not elaborated.

We remained uncertain whether the cap was in constant dollars, which provided increments for inflation, or "then year" dollars. A cap on constant dollar expenditure would still allow for significant annual increases in expenditures on the program. A "then year" dollar cap would be far more restrictive, in effect forcing increasing actual reductions as a result of the effect of inflation on the value of the dollar.

My own guess was that Rabin would apply a "then year" dollar cap, in effect squeezing the program to the point that it would be uneconomic. But he had no need to make his intentions clear at this stage. I felt, and I had no way to substantiate my instinct, that Rabin's ambiguity about the cap was another way for him to buy some time while adding to the pressure on the program. To my eyes, he wanted the Lavi to collapse of its own weight without having to be seen as the one who actually killed it.

The somber tone of our talks on the Lavi seemed to carry over to our discussions at the ministry about the navy program. The ministry's civilian staff stressed the very same financial constraints that we had tended to avoid in our previous discussions. I was unsure whether the staff concerns were genuine, or were merely a reflection of Director General David Ivri's sense that the Lavi was top-priority. In any event, the mildly negative tone threw something of a damper on our talks.

On our last day in Israel, Tuesday, June 3, I was the guest of the IDF on a tour of Israel's military facilities on the Golan Heights.

At my request, the chopper circled Temple Mount as we departed Jerusalem; I was no less thrilled this time than I had been the year before. Up in the Golan, we overflew an old synagogue that had been built in Roman times. I was fascinated to see that its layout was similar to that of a cathedral, complete with nave.

Once in the Golan, it was made very clear to us just how important the Heights were to the defense of the Galilee below. The edge of the Golan is like the redoubt that overlooks the Plains of Abraham below Quebec. The Golan offers a commanding position over the entire Galilee Valley and is seemingly impregnable. How the Israelis seized the Golan in 1967 always amazed me, just as I never could fathom just how Wolfe was able to seize Quebec two centuries before.

I was struck by the degree to which the Israelis had taken steps to ensure that any Syrian movement would be detected early enough to permit a call-up of the reserves. Moreover, if the Syrians penetrated into Israeli-held territory, their armored movements would be severely restricted by the combination of terrain and post-1967 infrastructure, making them easy prey for Israel's air force as well as ground troops.

At one point on our tour we joined General P. X. Kelley, the U.S. Marine Corps commandant, who was on an official visit as a guest of General Moshe Levy, the IDF commander. I was delighted to see P. X. I had already known P. X. for about ten years, and I still remembered when we had first met. P. X. had become director of the USMC Requirements and Programs Office and I was a junior analyst at the Congressional Budget Office. I was conducting a study of America's projection forces for the CBO, and I needed to get some marine perspectives. I went out to the old navy buildings on Columbia Pike, behind the Pentagon, which housed USMC headquarters. I spoke mostly to Kelley's staff and then spoke to the man himself. He seemed to tower over me. He also looked strong enough to lift me with one hand. I took an instant liking to the ruddy, talkative, and exceedingly bright hail-fellow-well-met marine.

At the time we met, in early 1978, P. X. was a two-star general on the way up. He knew Washington's ways and was a favorite with both the Hill and the analytical community. He came across as extremely outspoken, yet seemed untouchable, no matter what he said. Evidently Harold Brown took a liking to him as well. P. X. was appointed the first commander of the Rapid Deployment Force and soon was the darling of the press: America's swashbuckling marine hero who would make the Carter Doctrine a reality.[11]

Just before P. X. transferred from Washington to his new headquarters, a group of us threw a farewell luncheon in his honor. We were all certain that he would do a terrific job, and that he would someday be commandant, perhaps even chairman of the Joint Chiefs. P. X. seemed somewhat subdued—at least for him. Everybody knew that Jimmy Carter's foreign policy had been a disaster, and P. X. clearly was going to be a point man for the new White House foreign policy. We all also knew that in many ways the RDF was a bluff. As one former defense secretary had put it at the time, the RDF was neither rapid, nor deployable, nor a force. But P. X. was one of the shrewdest political operators in the military, or anywhere in Washington, for that matter, and he managed to turn his stint at the RDF into a stepping-stone for promotion to assistant commandant of the Marine Corps.

We kept in close touch when P. X. moved down to MacDill Air Force Base in Florida to run the RDF. Although his public persona throughout his tenure was that of a take-charge commander, he actually was very frustrated by what developed into a bureaucratic rivalry with the commander of the U.S. Readiness Command, who felt he was P. X.'s boss and saw Kelley as invading his turf. Once I was visiting P. X. at an exercise that some of his troops were having at Fort Bragg. When I walked into his tent, he handed me some Rolaids. "That's what this job is all about," he said. He then gave me a copy of a note he had received from his army boss, which talked about the murkiness of the chain of command and pretty much told P. X. to shape up. It wasn't all fun, games, and newsmagazine covers.

When I first came to the Pentagon, P. X. came around to my office. At the time I was a very junior official, and my office showed it; I worked in a windowless affair on an obscure "ring," as the hallways are called. P. X. was a four-star by then; my office mates were shocked to see him come in. No one higher than a full colonel ever bothered much with our "spaces." I asked him for a photo; "Everybody needs a general's photo when he moves here." Within a week I had one, complete with the most gushing prose. I gave his picture the place of honor on my wall.

P. X.'s rise to the top came to a screeching halt with the 1983 bombing of the marine barracks in Beirut. Until then many of his friends were convinced that he was a shoo-in to be the first marine to be chairman of the Joint Chiefs. The mass death took its toll on him personally as well as professionally. I saw him at the Marine Corps' birthday dinner not long after the tragedy. He was a shadow of his old self. He felt the losses very deeply, and it took him years to recover.

When he and I met in the Golan two years later, the marines had already weathered their confrontation with Israel's army in Beirut. Relations between the two military forces were excellent. In fact, the marines were acquiring Israeli Mastiff remotely piloted vehicles, and P. X. was visiting Israeli units partly to see what other systems might be worth acquiring for the Marine Corps. I recall that he was especially taken by an Israeli mine plow; the marines subsequently bought the system.

At one point during our joint tour of the installation, P. X., as was his wont, walked up to a group of young Israeli soldiers and asked one of them how he liked the service. "It's fine, *sir*," came the reply in unaccented English. "Where ya from, soldier?" "San Diego, *sir.*" "I've served in Pendleton. Do you know it?" "Yes, *sir.*" "Whaddya do here?" P. X. was given a brief overview of the soldier's duties. "Fine, as you were." "Thank you, *sir.*"

P. X. turned to me and said. "He's dedicated. They've got a good bunch here. They care about their country and what they're fighting for."

I've often thought about what Kelley said. There was no sign of rancor that a young man had chosen to live in Israel and not the United States. What counted was that he was prepared to fight for his adopted country and clearly respected the country of his birth. Unlike Jonathan Pollard, this young man did it the right way.

Moshe Levy and I chatted quite a bit during that trip to the Golan. I found him a bluff, likable fellow, in some ways an Israeli version of P. X. He invited me to visit his kibbutz on some future trip; I regret that I never took up the offer. On the other hand, we got to talking about Shamir, and I mentioned that my dad had asked me to look up his old friend the foreign minister when I was next in Israel. Levy said he could make that happen. Sure enough, on Tuesday evening, just after 9:00 P.M., prior to my departure, I went to Foreign Minister Yitzchak Shamir's home for a private visit. I was very pleasantly surprised that he had agreed to see me. The fact that I was to meet him in his home underscored the personal nature of the visit. He was not meeting with a DoD official; he was meeting my father's son.

Upon my arrival, Shamir met me at the door himself. His wife was in the kitchen; no one else was home. My first glance around the place left me with the feeling that I was in a typical Israeli middle-class flat. It was not in the least bit showy; it was almost spartan, in fact. A menorah, made of shells, stood prominently on display.

I had not realized how short Shamir was. I should have. Most of my father's neighbors, relatives, and friends from the old country were closer to five feet than to six. My mother often said that they were short because they were mostly poor and never really ate nutritious foods. Maybe Mom was right. Certainly the next generation, at least in the United States, appears to be a good head taller.

I also was struck by Shamir's ready smile. He had never received good press in the States, or in Israel for that matter, and not being particularly good-looking, was always being portrayed with a rather hangdog expression. Shamir made no bones about disliking the press. He said he preferred to ignore the media, and never let them bother him. His attitude did not hurt him all that much; despite constant sniping from Israel's bloodthirsty media community, Shamir managed to leave office in 1992 as Israel's longest-serving prime minister.

Mrs. Shamir briefly emerged from the kitchen to offer tea, fruit, and cake. Though she came across as a nice Jewish grandmother—a bubbie—which in fact she is (Shamir told me that he already had five grandchildren), there was something steely about her. I had heard that she had married Shamir while both were in the Jewish underground. Shamir told me later that evening that they had met in jail, in Djibouti. She must have been one tough lady in her youth.

We—or rather he—spent some time reminiscing about the old days. My father had recalled that Shamir was a very sentimental sort, but I had not realized just how sentimental the man was. He told me that my grandfather, who had been liquidated by the Nazis, was a tall man; that my uncle was a "wild" boy. I was amused to hear that description of my dad's brother, as proper and formal an Orthodox rabbi as ever I have come across. My dad, he said, was always an energetic sort—which was still true a half century later. Shamir had been a classmate of one of my aunts, my dad's older sister; he lived near my family. He had emigrated to Israel and joined the underground, and was arrested and jailed by the British in Eritrea, and was then transferred to Djibouti.

Shamir asked me a bit about my personal background, but soon reverted to talking about his own, as well as evaluating the many people he met and knew. He talked about his days in Paris, as the Mossad's representative there. He would close his eyes and sigh; he really loves the place. He pointed proudly to a little statuette of the three monkeys—hear no evil, see no evil, speak no evil—and said that his Mossad colleagues had presented it to him when he left. Even after so many years, he seemed really taken with that small gift.

He then took me on a whirlwind tour around his political world. Sadat, he said, "was a great man," a rather surprising comment from someone who opposed the Camp David Accords. Even more surprising was his characterization of then Prime Minister Andreas Papandreou of Greece. "He isn't so bad," he said about the man who seemed to caress the PLO and constantly stick it in America's eye. He admired Margaret Thatcher and thought that Prime Minister Felipe González of Spain had "changed" when he came to power. The Spaniard had established relations with Israel, helping to close a wound with the Jewish people that had festered for half a millennium. "The United States helped on that," Shamir said. On the other hand, Turkey was a problem. Israel had once had excellent relations with Ankara, but these had deteriorated after 1973. He felt that the "Jewish lobby" helped with Turkey, as did the lobby's relations with its Greek counterpart in Washington. Shamir stressed that Europe in general was anti-American, regardless of the lip service it paid to Washington. One thing was certain, though: Europe in the mid-1980s was not very pro-Israel.

I asked Shamir about Africa, but he was more interested in discussing East Asia. That, in his view, was where Israel's future political and economic prospects lay. There were problems with the Japanese. Shamir felt that Shintaro Abe, then Japan's foreign minister, was particularly hostile to Israel, while Prime Minister Yasuhiro Nakasone was more disposed to improving relations. Israel maintained a consulate in Hong Kong as a window to China. Shamir said that the Chinese were pressing for more business. We knew about Chinese-Israeli ties, but I was a bit surprised at Shamir's openness about them. When he became prime minister he designated Arens as a key point man on China, reflecting the growing military ties between the two states. ← *Sold extremely sensitive and*

Our discussion covered many other areas as well—South Africa, Ethiopia, submarines, Nicaragua. We did not discuss the Lavi. Shamir clearly saw the evening as a social one, as a way to get to know his friend's son—he penned a personal note to my father the next day—

extremely secret military information to
China —

and he was not going to spoil it by injecting an issue that brought me into conflict with the man he called his "good friend," Moshe Arens.

I left Shamir's house nearly two hours after I arrived. I hadn't expected more than a perfunctory conversation. I came away with the feeling that people in Washington seriously underrated Mr. Yitzchak Shamir.

11

The Debate Heats Up

I never get much sleep on the last day of a visit to Israel. Though I usually flew into Israel via air force jet from Europe, I flew directly home, and Pentagon regulations required that I fly with an American flag carrier. All flights to the States normally depart very early in the morning, and we would have to leave the hotel for Ben Gurion Airport around 3:30 A.M.

The combination of adrenaline overload and the early departure kept me awake as usual. Watching television after I returned home from the Shamirs' flat, I saw the report of George Shultz's supposedly secret letter to Prime Minister Peres, which had been timed to arrive when our team was in Israel. The televised report showed a copy of the original letter, together with a translation in Hebrew. The gist of the note was that it was essential that we and the Israelis bridge the cost gap over the Lavi for the plane to receive continued support from the United States. This, of course, we had not done.

I was amused somewhat later to hear Menachem Eini tell me that I had drafted Shultz's letter as well as the letter from Cap Weinberger that Tom Pickering had handed to Rabin during our meeting. While I had been very much involved in the drafting of the Weinberger message, I barely came into contact with George Shultz on the Lavi or any other issue. That Eini could think I was so sinister and manipulative was no small thing, however, since it added to my credibility when I claimed to speak for the administration's most senior officials.

Whatever Shultz's underlying motives might have been, his message was seen as reinforcing that of Weinberger, and it clearly chilled the Israelis. The Israeli press in particular correctly interpreted his intervention as one that had major long-term significance for the survival of the program. Unlike Cap Weinberger, Shultz was considered to be Israel's friend. More than that, the Israelis viewed Shultz as the most friendly occupant of the seventh floor of the State Department since the creation of the State of Israel, Henry Kissinger included. Anything that Shultz said or did was virtually gospel in Israel. For Shultz even to hint at wavering on the Lavi came as a shock to the plane's supporters.

I was able to get a clear sense of Israeli apprehensions while I flew back to Washington. The news kiosk at Ben Gurion had opened just minutes before we were due to board and I was able to pick up a copy of the *Jerusalem Post*. The front-page headlines told all: "Doubts About Lavi Fighter Grow: Zakheim Meeting 'Dismal Failure.'"[1] The *Post* noted the increasing despair of Israeli officials who had somehow hoped that we could be cajoled or bullied into accepting their estimates and now were "upset" that the costing gap was unlikely to be closed.

I could not understand then, nor do I understand with nearly a decade's hindsight, why the MoD's Lavi team expected a sudden change on our part. The *Post* story indicated that the major differences between my team's figures and those prepared under Zvi Tropp were the costs of the engines and the labor rates. These differences, and many others, had been the basis for the Israeli note that I had received a few weeks earlier, whose calculations our team could not accept. We had given no indications that we accepted the premises on which their evaluation was based. Zvi in particular knew how strongly I felt about the validity of our numbers. He had often broadly hinted his concern about the reliability of the Lavi team's calculations, though he had never actually repudiated them. Perhaps the belief that I would recite a form of *mea culpa* when I met with Rabin was a manifestation of the panic that was setting in. The *Post* rightly recognized that if Shultz was holding the project hostage to an agreement on costs, and if we were not yielding on the figures, the Lavi was in even more trouble than anyone in Israel had realized or expected.

Whatever doubts Shultz may have harbored on the true costs of the plane, he was still far more sympathetic to the program than Weinberger and, for once apparently unbeknownst to the Israelis, was at odds with the Pentagon over the issue of Lavi contracts. The fact that the Israelis sought to speed up the Defense Security Assistance Agency's approvals of Lavi-related contracts continued to annoy Weinberger, who saw the exercise as one of pouring good money after bad. Shultz, on the other hand, was still ambivalent on the issue despite the message to Israel. He was considerably more reluctant to do anything to slow down Israel's ability to draw upon our security assistance in support of Lavi contracts. As was often the case between the two men, neither was prepared to give any ground on the issue. Ambassador Tom Pickering had easily won State's support for the study-for-contracts arrangement that Rabin had suggested at our meeting with him. I was having a lot more trouble with Weinberger. I could not get through to him on the matter at all.

For several weeks in June 1986 the two principals were dead-locked, though even Weinberger was not prepared to issue a blanket hold on all the contracts. In fact, it was difficult to get the two men to focus on the issue at their weekly breakfast meetings, despite Fred Iklé's best efforts to prompt the SecDef on the issue. Cap Weinberger had other, far more important matters to resolve with Shultz, and he let the Lavi matter drift. And Shultz certainly was not going to cave preemptively to Weinberger.

Even as the two secretaries reached no agreement, however, we continued to put a stall on Lavi contracts. We were certain that Weinberger wanted to block them all and thought that Shultz ultimately would yield to Defense's position. In any event, Weinberger finally moved unilaterally on June 23, and we braced for the latest congressional onslaught as word would leak out that we were not permitting any contracts to be implemented.

In the meantime, the debate inside Israel became more pronounced virtually as soon as we departed Israeli soil. Within two days, both Rabin and Peres acknowledged that Israel's own estimates left something to be desired. The left-leaning *Al Hamishmar* reported that Rabin felt that the Israeli figures "lacked consistency" and were "not 100 percent accurate." The paper added that as a consequence of our team's visit, Rabin had given the Lavi team a month to resolve what appeared to be contradictions in their own calculations.[2] Even Peres, that stalwart supporter of the high-technology project, conceded to the *New York Times* that the Lavi "will cost less than the Pentagon thinks and more than our people think."

Neither man was intimating that the program deserved to be canceled. Nor had either yet breathed a word about alternatives. In fact, the *Times* story concluded by observing that whatever their other differences, both the Likud and Labor factions of the National Unity government supported the plane and that, despite growing criticism, "the betting here, even among Israeli critics of the Lavi, is that the project will survive because to kill it outright would be too damaging to Israeli-United States relations."[3]

The *Times* had overlooked the significance of what Peres had said, however. Even the admission that the Lavi's numbers were flawed was a major blow to Eini and the IAI team. It was bad enough that Shultz was insisting that the Israelis climb down from their position on costs; now even Peres and Rabin were doing just that. But if there was to be a compromise on the estimates, and we, for our part, were not budging, just how were the Israelis to alter their projections? To have yielded on the issue of either wage rates or engine costs would have instantly shattered their credibility and would have had serious implications for the production costs associated with the plane. Since the development cycle was nearing its end, the real issue for the Israeli military was whether the production run was going to consume more than the Lavi's proponents predicted. If it did, other programs really would be in as much trouble as we were saying they were.

The Lavi team was in a tight spot, and the Israeli military did not make it any easier for them. For the very reasons that gave the Lavi team pause, the IDF leadership felt compelled to be less circumspect about the importance of the program to Israel's future security needs. In a surprising statement, Amos Lapidot, the air force commander, who had always staunchly backed the plane, bluntly informed the New York Hebrew-language paper *Yisrael Shelanu* that the Lavi was not vital to Israel's defense. He added that Israel could consider other types of aircraft for its air force, and that Israel was unable to support the project alone, "either financially or technologically."[4] Lapidot's interview constituted the first public acknowledgment of Israel's willingness to listen to what we had to say about options other than the Lavi. Not surprisingly, other generals, less identified with the Lavi to begin with, openly took up Lapidot's cue. A week later, the same newspaper quoted another Israeli general that "if the Lavi takes off, the army will be prone on the ground."[5]

Making matters even worse for the Lavi team, there appeared for the first time some public chinks in the congressional wall of support for the project. Speaking to reporters in Israel, Senator John Kerry, a member of the staunchly pro-Israel Foreign Relations Committee, stated that if there was no agreement between the United States and Israel on the cost of the plane, there might be a cut in the level of aid to Israel.[6] For Lavi proponents, Kerry's pronouncement was the most worrying of all. Nothing creates as much fear in the hearts of Israel's ruling elite as a threat to cut American financial assistance. Kerry's statement appeared to make that assistance a hostage to the Pentagon's estimates, since we were not about to change our position on the plane's costs. The Lavi's supporters had to share one conclusion with us: the tide in both Israel and Washington was beginning to turn against the plane.

The response of the Lavi camp was to circle the wagons. The accusations of a Machiavellian plot on the part of the Pentagon to crash the Lavi became increasingly shrill and more obviously orchestrated. The Lavi was being threatened by a sinister alliance of industrialists, anti-Semites, self-hating Jews, and Arabists. The pressure on Israel was being applied at every possible opportunity, the goal being to undermine the plane both in Israel and in Washington.

The description of a meeting that I had held with a group of members of the Israeli Knesset not long before our trip to Israel provided a case in point. The MKs (as they are termed) were visiting Washington as part of a monthlong tour of the United States. The Pentagon was included on their itinerary, and I was asked to address them. Naturally enough, the Lavi was the major theme of the meeting. Some of the group clearly had doubts about the project, among them Haim Ramon, who would later rise to prominence as one of the architects of the 1992 Labor victory, midwife of the Israeli-PLO accord, secretary general of the Histadrut labor federation, and Minister in the Peres government. Others, primarily from the Likud, made no effort to conceal their hostility to what I had to say. Their questions had a cutting, harsh tone that made it clear that I was no favorite of theirs. Upon their return to Israel, just two days after I departed for the States, they informed the press and Israeli radio that they had been subjected to a concerted effort of what the radio termed "gentle pressure" designed to turn them against the plane and that what really was at issue was American industrial interests.[7] At about the same time, *Defense News* reported the claim of unnamed "Israeli industrialists in Washington" (probably IAI's Washington office) that rumors of an imminent cancellation of the project were part of "an orchestrated program by elements in the Pentagon to discredit the Lavi."[8] If anything was orchestrated, it was the virtual identity of language employed by the Lavi's Knesset supporters in Jerusalem and her IAI supporters in Washington.

We followed the debate in Israel with caution, but strenuously avoided any public comment about the plane's future. None of us was foolhardy enough to contemplate openly the cancellation of the program. To have done so would have been to undermine the viability and credibility of the alternatives study before it was even written.

Yitzchak Rabin, with his finger carefully on the pulse of public debate, was not as blunt about the plane as his military officers. He did tip his hand a bit further in our direction when he stated in a June 10 interview to *Al Hamishmar* that "other alternatives which the Americans wish to offer, including U.S.-made aircraft, are certainly worth considering, provided the Lavi project is not put off."[9] Rabin expanded upon his point in response to a question in the Knesset the following day:

> Elements in the U.S. Administration never tried to order us to stop developing and producing the Lavi. I think that such a thing does not exist in relations between the two countries. They are only trying to convince us on a change of direction regarding the Lavi's production, not so much regarding its development. At the same time, however, if elements in the U.S. Administration raise alternative solutions for our two problems—a plane tailored to meet our operational needs and the continued development and strengthening of a technological-industrial infrastructure—in my opinion we should listen to these proposals and make our own decisions. All we have agreed to is to examine the alternatives, if alternatives are raised.[10]

Rabin was not dissimulating. He had every reason to assume that we would not block Lavi development, since the development program was at least half complete, and since he controlled the timetable regarding any decision on the aircraft. Moreover, Rabin was committed to strengthening Israel's industrial base every bit as much as Peres. He correctly saw that even if Israel did not have the industrial and financial wherewithal to produce an airframe that was more cost-effective than an American equivalent, it needed to maximize its comparative advantage in producing high-quality and cost-competitive avionics.

Rabin's remarks also laid out publicly for the first time the two key criteria for choosing an alternative, and around which our study was to be constructed. First and foremost, Israel had to be convinced that it could obtain the same operational capabilities from an American aircraft as from the Lavi, if not better. On that score we had little doubt that the Israeli air force could be satisfied. Second, however, Rabin also had to placate his industries, since getting them development work without any production jobs not only would render meaningless their avionics work but would result in widespread layoffs that would disrupt their highly skilled workforce. We believed that we could satisfy Rabin on the work shares issue as well, but we were less sure of ourselves and very much dependent on the defense contractor community for information and proposals. Still, Rabin's remarks were an important signal not only of his open-mindedness about the issue but of his reliability. He made public what he had told me in private five weeks earlier—he was, as always, a man of his word.

Not all of the Israeli press, and certainly the Lavi's proponents, read Rabin's statement as we did. As the *Jerusalem Post* advertised in its headline, "Rabin: Lavi Project Will Go On."[11] Moshe Arens, ever on the lookout for the welfare of his pet project, responded to the Rabin statement by stating that "we're very happy to listen to any advice, for or against, but the decision is ours, the responsibility is ours, and the decision has been taken."[12] Of course, Arens bitterly resented our advice, and, as I was beginning to learn, resented me personally, though we had met but once since I had undertaken the project, and never at all while he served in Washington.

The *Jerusalem Post*, in an editorial, picked up on another strand of Rabin's argument, again in order to justify the case for the Lavi:

The dialogue with the U.S. about the plane cannot . . . be limited to differing assessments of production costs. . . . The debate over the Lavi, if it is to be genuine, must shift away from cost overruns to Israel's technological future. And the real question then would be whether there is an alternative project or projects that would give Israel the benefits that its industrial and technical base can derive from the Lavi program. . . . From that perspective, it might even be concluded that having gone so far already with the Lavi it remains the best alternative.[13]

The *Post*'s argument would later be echoed again and again as the fate of the Lavi hung in the balance. The notion that somehow more good money should be poured after bad—because of the costs that have already been sunk—is one that invariably is articulated by weapons manufacturers everywhere as they confront the reality that their programs are truly on the chopping block. It is the military-industrial equivalent of whistling in the wind.

Still, the *Post* represented just one side of the high-profile debate. Writing on the same day as the *Post* editorial appeared, defense correspondent Reuven Pedatzur argued against the project in a long *Ha'aretz* commentary that emphasized that "the aircraft will leave Israel without a single dollar for the purchase of new weapons systems."[14] Similarly, *Ma'ariv* argued in an editorial that the issue was not whether the Lavi would be produced, but rather what was the best alternative to its production.

Our embassy in Tel Aviv interpreted the commentary surrounding Rabin's statement in much the same fashion as we did. The staff noted increased public pressure on Lavi proponents to justify their cause. They also shared our view that Rabin was in no way backtracking from his agreement to participate in an alternatives study.

Interestingly, Embassy Tel Aviv saw a connection between the Lavi dispute and the Pollard affair. Some of the staff saw Israeli behavior on the Lavi as one of preemption, as one embassy official anonymously put it in unbalanced couplet form, "attempting to rally around the 'blue and white' as things are starting to get tight." The *New York Times* had seen a similar connection, though from a slightly different perspective. Its June 4 story on the Lavi reported that "there is much anxiety in this country over the possible damage to Israeli-United States relations as a result of the Pollard spy affair." Perhaps that is why it concluded that the Lavi ultimately would fly: neither the United States nor Israel wanted the friction engendered by Pollard to spill over into the Lavi dispute and eat away at what was an otherwise healthy relationship.

Publicly, Ambassador Tom Pickering continued to go to great

lengths to reassure the Israelis that not only was the relationship healthy, but that we were not out to kill the plane. In a widely reported "private" speech to the Anti-Defamation League, Pickering denied unequivocally that we were seeking the program's termination. He insisted that we were only trying to suggest economically feasible alternatives to the Lavi. Semipublic assurances notwithstanding, the embassy staff privately made no bones about where it stood on the Lavi. As the *New York Times* reported, "sources at the American embassy here have been putting out the word that the United States would like to kill the project."[15]

Our embassy's mixed signals on the Lavi further increased IAI and MoD anxieties, as, no doubt, did Rabin's increasingly lukewarm support for the project. The response of the plane's producers was to market the plane *outside* Israel. We did not believe for a moment that any nation other than Israel would be seriously interested in the fighter. To begin with, we were certain that the Lavi, with its small production run, would price itself out of the international market, much as Swedish aircraft had done over the previous two decades. Moreover, for the Lavi to be sold abroad would require the U.S. government's approval, since so much of the plane was produced in the United States. The Israelis were having trouble getting their far less sophisticated Kfir approved for sale to Latin America; the likelihood that the Lavi would win such approval was minimal. Finally, the Israelis' insistence that the Lavi was specially designed for their air force's unique needs—Chief of Staff Moshe Levy said that it was "like having a suit tailor-made or picking one out at a store"[16]— implied that it would be a less than efficient solution for another state's requirements. Surely a supposedly more general-purpose aircraft, such as the F-16, would be more suitable for a state that did not have operational needs identical with those of Israel.

None of the foregoing deterred the IAI marketeers, of course. It was not long after our study appeared that we began to hear rumors of IAI representatives trying to sell the Lavi in the company's traditional hunting grounds, notably in Latin America. Late in April we obtained evidence on this score that was considerably more concrete. Fred Iklé was sent a copy of a full-page advertisement for the Lavi that IAI had placed in *Fuerza Aerea*, the Chilean air force's official publication. The page carried a picture of the plane in its hangar, with the caption *"Avion Lavi, fabricado por la Industria Aeronautica de Israel."*

Both Iklé and I found the ad infuriating. It not only conveyed the impression that the Lavi was for sale overseas—which was impossible without our consent—it also gave the lie to the Israelis' claim about its uniqueness to the Middle Eastern context, which on its face was the most telling *operational* case for the plane.

The White House was also caught by surprise, and made no bones about its annoyance. *Aviation Week* quoted a "White House official" who asserted that "we were told categorically that the Lavi was not being developed for export. Now they want to export to the U.S. and third countries."[17] The journal also reported that American officials had concluded that the Israeli cost estimates presupposed export sales of up to two hundred aircraft. In fact, we had not explicitly attempted to reconstruct the Israeli estimates. But it was clear to us that unless Israel exported a significant number of aircraft, production runs would be too low to enable IAI to come even close to meeting Israel's stated cost targets.

It was good, from the DoD perspective, that the White House was being quoted on the issue of exports, and that the Israelis saw that other agencies shared our concern about the long-term prospects for the airplane. The Israeli response to the White House concern was a bit too cute: Israel had no *present* intention to export the Lavi. "With U.S. permission to re-export U.S.-made parts, export sales might be a possibility," an official told *Aviation Week*.[18] It was not easy to figure out how this response squared with the advertisement in the Chilean air force magazine. After dealing with the Israelis for over a year, however, nothing surprised me anymore.

12

Enter the GAO

Throughout the late spring of 1986 the General Accounting Office study was only beginning to take shape, even though we were getting into full swing on our alternatives project. It became clear that the GAO would report its findings around the time that our alternatives study would be completed. The Israelis made every effort to present their best case to the GAO staff. Everyone on both sides of the issue recognized that the DoD follow-on effort, which would dare to compare Israel's precious system to American planes that already were operational, was likely to be even more disruptive to the program than its initial study. The Israelis hoped that if the GAO repudiated our estimates, not only would our initial figures be rejected by the Congress, but the credibility of our alternatives report would also be seriously damaged. Lee Hamilton, who chaired the Middle East Subcommittee of the House Foreign Affairs Committee, and who had requested the report, was seeking something far more straightforward. As one of his staffers put it shortly after we returned from our trip in July, "With all the haranguing going on and the cost projections going up and up, it does raise questions about what's going on in the program."[1]

Not long after the GAO had begun its effort, we had met in my office with Al Huntington, the GAO's project officer, and his colleagues. Al had outlined his goals for the study, and we offered to be of any assistance that he and his staff desired. Throughout the GAO

159

effort we were completely forthcoming with Huntington, a likable and straightforward analyst almost exactly my age, who seemed quite capable of carrying out his task effectively. Dick Smull kept in close touch with him as the GAO pursued its factual trail. He and his team also met with people from State, as well as from the Israeli embassy and Ben-Joseph's purchasing team in New York. The Israelis urged the GAO analysts to visit Israel, in order to get a firsthand report from the Lavi teams in IAI and the ministry of defense.

GAO analysts traditionally rely on their own eyes and ears as much as on documentation; study trips are the norm for all major, and many minor, GAO projects. So it came as no surprise to us that shortly after our arrival back in the States, the GAO requested State's support for a fact-finding trip to Israel by Huntington and his deputy, to take place during the second and third weeks of July. Trips overseas by officials of the U.S. government are generally "at the discretion of the ambassador" to the country in question. Meetings with foreign officials are invariably arranged by embassy staff. Without State's support, the GAO trip would have been a failure, or maybe would not have taken place at all.

State generally does not rubber-stamp trip requests put to it by Congressional staff. On occasion it actively discourages visits to a particular country. Sometimes the embassy staff is deliberately unco-operative if the visitors make the trip against the advice of headquar-ters. Trips coordinated with an embassy usually result in embassy staff demanding to accompany the visitors on official meetings with foreign government officials. At times the visiting Congressional staff avoid or evade such escorts, causing much irritation at the embassy and prompting nasty cables back to State.

State had no problems with this particular visit, however. The department had little desire to mess with the GAO. The wrath of the Congressional auditors was known to make federal departments very uncomfortable, and State diplomatically elected to leave well enough alone. The cable from headquarters to Embassy Tel Aviv made State's preferences plain to any embassy personnel who might be less than enthusiastic about cooperating with the fearsome agency's rep-resentatives: "It is the department's policy to be of assistance to the GAO and to be as helpful as possible." Of course, the cable added that the appointments were to be made through the embassy and that the embassy would provide escorts to the meetings. State was prepared to cooperate; it was not prepared to give away its traditional turf.

The GAO request to State outlined the direction that the study was taking and the progress that the GAO felt it had already made. First of all, the study team intended to focus on the history of the Lavi program and the changes to the aircraft design requirements since the program's inception. From the Israeli perspective, this part of the study was unlikely to yield results friendly to the project. The Lavi had evolved from a low-cost fighter to an expensive competitor to the F-16. That evolution, as so many observers had noted, had taken place virtually without the detailed oversight of the ministry of defense. It had resulted in the plane's increased costs, and it begged the question of just why the Israelis had elected to jettison their original concept for the aircraft.

The GAO intended to evaluate the impact of those design changes on both the plane's cost and production schedules. The team also wanted to discuss potential future design changes, the impact of these on the cost and production schedules, and the provision the Israelis had made for such changes at different stages of development. Our study had made only minor allowances for such changes, and indicated for that reason that our estimates were conservative. If the GAO really pursued this theme, it could open an even bigger can of worms than we already had.

The GAO also felt that it as yet had incomplete information regarding the disbursement of American funds for the program. The Israelis promised to make that information available "in country." It was not clear to us why Israel simply could not transmit those figures through its attachés in Washington or its New York purchasing office. We suspected that, Israeli promises notwithstanding, the GAO would find little more in Israel than it had already obtained in Washington.

Huntington and his team requested an Israeli evaluation of the differences between the two sets of estimates, particularly relating to labor rates, engine costs, and material costs, and the Israelis happily seized upon the opportunity to rebut our findings. On the other hand, the Israelis could only have been uncomfortable with some of the more pointed GAO requests to discuss:

- whether Rabin's $550 million funding cap was a yearly average or a ceiling for any given year;
- the impact of the cap on the number of aircraft that were to be produced;
- anticipated production problems and the impact of delays on the cost estimates;

- Israel's plans for component production, and the impact on the cost estimates if components were acquired from sources (including foreign countries) other than those initially anticipated by the ministry of defense; and
- the impact of the Lavi on the availability of funds for other priority Israeli programs.

In addition, and most awkward of all, Huntington wanted to explore the issue of Lavi export sales. He informed the Israelis that he intended to discuss their expectations for such sales; efforts they had already undertaken to market the plane; their expectations of sales to the United States; and their view of Israel's prospects for obtaining American permission to export the plane, since so many of its most important components were American. I would have loved to have been in the room when Moshe Arens was informed that these last issues were also on the GAO agenda.

Huntington and his colleague did not receive all the information they sought. The Israelis saw the report's value in terms that were quite different from those of the two analysts, and the MoD and IAI fed the Huntington team only such material as they deemed would strengthen their case for the aircraft. We learned from Huntington, and later read in the report itself, that the Israelis were very circumspect about the details of their wraparound rates, and even more tight-lipped about their calculations regarding future cost increases and their plans for exporting the Lavi.

During the summer we were able to obtain a GAO document entitled "Preliminary Conclusions on Bridging the MoD & DoD Cost Estimates." The top and bottom of every one of the document's five pages was marked DRAFT in large capital letters. The GAO's logo or name appeared nowhere in the document.

The unclassified paper opened with a statement of the two flyaway cost estimates: $15.2 million for the MoD, $22.1 million for the DoD. It then summarized in two columns under the headings "MoD Estimate" and "DoD Estimate" where additions and subtractions needed to be made. It shifted certain cost estimates in the Israeli presentation from the research and development category to that of production, resulting in potentially higher future costs, since production had not yet started. At the same time, the paper reduced the DoD development cost estimate by nearly $700 million. When it came to procurement costs, however, the draft paper added $1 billion to the Israeli totals while also deducting about $1.5 billion from our estimate. Finally, the GAO included a table that showed that annual inflation-driven increases at rates as low as 7.5 percent would double the dollar cost of the Israeli program and would result in outlays on the Lavi that would approximate $1 billion every year from 1991 to 2001.

This last chart was devastating. For although the GAO had adjusted our estimates more than those of Israel, it also indicated quite clearly that in the absence of increases in security assistance funds to Israel, the Lavi would virtually wipe out all other Israeli program initiatives throughout the 1990s. A section entitled "Bottom Line" concluded that "regardless of the adjustments, MoD's cash flow estimate, accounting for modest inflation (5 percent), exceeds Israel's expenditure cap ($550 million) in 1989 and rises to more than $900 million in years 1991 to 2000."

The Israelis were dismayed by the team's preliminary conclusions, and Huntington soon found himself in an uncomfortable position. Through friendly members of Congress, the Israelis did their best to apply whatever political pressure they could on the GAO. We learned that Huntington's superiors had decided not to let the analysts' unvarnished conclusions see the light of day.

We harbored no illusions about the GAO. The DoD and the GAO were so frequently at odds over cost issues that we were prepared to welcome any degree of agreement between the two agencies. We also hoped to receive some recognition and acknowledgment by the congressional agency not only that our work had been conducted in a thoroughly professional manner, but also that the agency could not dismiss our findings, even if it did not fully subscribe to them.

That the GAO had initially validated our key finding, which was not what the plane actually might cost but rather whether it was affordable given the availability of Israeli and American resources to support the project, was a more than pleasant surprise. We knew that it would be a close call as to whether this conclusion would be reached, or even whether our more modest expectations regarding the professionalism of our effort would be fulfilled. A half year was still to elapse before the GAO reported; a half year for Israel to continue its high-pressure drive for the GAO to issue a report that would completely destroy our credibility.

13

Rollout

My latest visit to Israel, and the inability of the Pentagon and MoD teams to reach a consensus on the Lavi's costs, spawned a new reexamination by the Israeli defense ministry of its cost estimates. I did not expect to see much come out of the new evaluation. The Israelis had not really moved very far from their initial position and seemed to be conducting this latest exercise as a run-up to the rollout of the plane, which was now scheduled for Monday, July 21, 1986.

To show that the government remained foursquare behind the plane, even as the reassessment was being undertaken, Rabin announced to a businessmen's audience that the "inner cabinet," which made the government's most important decisions, had reaffirmed its support of the program by an 8–2 vote. We were convinced that, as before, the two opponents were Ezer Weitzmann, the true father of the Lavi, and Yitzchak Modai, the finance minister. More important, perhaps, the same report that carried Rabin's comments also relayed the pessimism of a number of MoD officials about the program's real future.

Whatever the feelings inside MoD, however, the PR drumbeat continued. Israel launched a new effort to gauge U.S. government support for the project. Jerusalem reportedly was going to "concentrate on assessing what power the Pentagon has within the administration on this issue, and what congressional support can be expected, given the atmosphere of budgetary constraint in Washington."[1] The rollout fit neatly into this strategy. The Israelis had invited a large congressional delegation to attend the affair, and we expected a high turnout among the invitees.

One particular incident illustrated not only the growing importance of the Lavi issue on the Hill but also more generally the clout that Israel's supporters wielded in town, and particularly their ability to monitor, react to, and influence developments in the U.S. government in virtual "real time." One day late in June, Fred Iklé and I were asked to brief Senator Bob Kasten on the status of our efforts with respect to the Lavi. When I was informed of Kasten's request, I was led to believe that it would address our plans for the alternatives study. Kasten was chairman of the Senate Appropriations Committee's Subcommittee on Foreign Operations. It was his subcommittee that marked up the security assistance bill, which was the source of major American financial support for Israel. His reaction to our work was therefore critical to the success of the effort. I had known Bob for a few years, and had found him an especially genial individual. He tended to align himself with the conservative wing of the Republican Party and was a strong supporter of President Reagan and his policies.

Given his role in the Senate, Kasten was always assiduously courted by AIPAC, but he was personally a staunch supporter of Israel as well. So too was his chief staffer on the subcommittee, Jim Bond, with whom Moshe Arens had worked so closely while ambassador to Washington.

Kasten's request for a briefing came at relatively short notice, but we dutifully made the fifteen-minute trip from the Pentagon to the Hill. Upon arriving I was stunned to learn that ninety minutes earlier—or about an hour after our meeting was scheduled, Danny Halperin, the economics minister at the Israeli embassy, had placed a call to Jim Bond to discuss the meeting. Danny was in Jerusalem when he made the call to Jim. He must have learned about our meeting within minutes of its having been scheduled. I felt as if we had an Israeli "big brother" watching us every time we did anything relating to the Lavi and the Hill.

Kasten indicated to us that he had no objections to an alternatives study per se, as long as the Israeli government had no objection. He stressed, however, that once the government came to a decision on the fate of the plane, it should do so without pressure from the Pentagon or anywhere else. He added that he did not see the alternatives study, and Israel's acceptance of the exercise, as some sort of justification or validation of our cost estimates. The alternatives study stood on its own; the issue of the estimates would be dealt with by the GAO, which he viewed as an impartial source.

Kasten then said that we should not withhold any Lavi contracts, since that would be construed by the government of Israel as a form of pressure on our part. Since Bob had been one of the signatories of the first letter to Weinberger regarding composites for the Lavi—the one that had not been sent because Weinberger had preempted the issue by releasing the contracts—we knew he was deadly serious about this matter.

Kasten concluded by saying that he thought we had not been as forthcoming with the Hill as we should have been, and suggested that we brief several of Israel's strongest supporters. These included Senator Dan Inouye, who was ranking minority member of the Foreign Operations Subcommittee, and with whom Kasten worked closely on Mideast issues, including the Lavi—he too was a signatory of the letter on Lavi composites. Kasten also suggested yet another cosigner of that nonletter, Jack Kemp, and Charlie Wilson, the father of the offshore procurement amendment.

Bob Kasten made his points in a very nonthreatening way. Jim Bond was not as gentle when he phoned me the following week. He had not liked our presentation at all, and he made it very clear to me that he, for one, and his senator were not backing away from their unswerving support for the program.

My conversation with Jim Bond was soon followed by the arrival of a letter to Fred Iklé from Kasten, dated June 27, the day after our briefing. I presume it was drafted by Jim Bond. The letter reiterated Kasten's points to us, for the record as it were. I was struck by his statement that he did "not have any objections to the alternatives to the Lavi program being examined should that be the wish of the Israeli government." That left little doubt in my mind who was calling the shots on this issue. As if to ensure that we not miss the obvious point, Kasten concluded his note by stating that he was not only passing copies to Inouye, Wilson, and Kemp, but was "making a copy of this letter available to the government of Israel."[2]

There seemed to be no letup to the pressures engendered by the Lavi, and I was delighted to take a few days off for the July Fourth weekend.

We had been invited to the festivities in New York Harbor to commemorate the Statue of Liberty Centennial, and I had a chance to spend some time with my parents, for whom I had been able to obtain tickets to the ceremony on the aircraft carrier *John F. Kennedy* on the morning of the Fourth.

We began our weekend by attending an exciting New York Mets game at Shea Stadium against the Houston Astros. The Mets were looking like champions, which they ultimately would be. (That was the year that the curse of the Babe struck again when the Red Sox' Bill Buckner let a ground ball go through his legs in the sixth game of the World Series.) I hadn't been to Shea in many years, though I had been a Mets fan since 1962. I had often regaled my kids with remembrances of Mets' games past. I had first seen them against the Dodgers, who were back in New York for the first time since having betrayed my love and that of everybody else in Brooklyn. It was Memorial Day, 1962, and the Mets were playing in the Polo Grounds. Brooklyn's longtime first baseman Gil Hodges, now a Met, hit three homers that day. His teammate Elio Chacon (remember him? He played shortstop for the Mets) started a triple play by spearing a line drive he should never have caught. Sandy Koufax gave up five runs in the first game. The Mets lost both games anyway.

I had been to Shea Stadium in 1964, the year it opened, and over the years had seen many of baseball's greats there. I had even seen Jim Bunning of the Phillies, now Congressman Bunning, Republican of Kentucky, pitch a perfect game against the Mets. So I had lots of stories to pass along to my kids, who had never been to Shea, and who had been brought up on a steady diet of Robinsons, Ripkens, and other denizens of Baltimore's Memorial Stadium.

The Mets game was as exciting as they come: they won in the bottom of the ninth on a homer by Darryl Strawberry, whose personal problems were still very much in the future. It was the perfect prelude to the Fourth itself, which was John Lehman's show. John has always had excellent connections with the Higher-Ups, and they did not fail him on that Friday. The weather could not have been better, and New York Harbor was a picture postcard full of small sailboats and outboards that surrounded the aircraft carrier and the tall ships that made their way into the harbor from Europe and South America.

We were seated behind Carl Trost, the chief of naval operations, in the second row of chairs on the flight deck facing the statue. The kids loved it, though I was a bit concerned that Saadya, in his enthusiasm, might knock over the CNO or one of the other notables seated alongside him. The boys had the best of all worlds: second-row seats, lots of military ceremonies, famous people whom they didn't really know much about but knew were "famous," and two doting grandparents who seemed ready to buy them any souvenir they wanted.

John Lehman's arrival on deck was suitably flamboyant, and added to the kids' thrills. Instead of crossing the water from Brooklyn by launch with the rest of the official party and guests, John came aboard by helicopter. My kids' eyes popped as the chopper landed on deck and John got off and shook their hands.

The navy secretary did not get my mom's special attention, though. Walking along the deck we met Mayor Ed Koch, who bussed her on the cheek. A typical New Yorker, Mom approved of Koch's blunt style. She had never personally met the man, and when he kissed her, she looked as if she were about to swoon, the way she told me she did when she had worn bobby sox at Frank Sinatra's early concerts. I teased her that she wouldn't wash that cheek for a long time.

Nice as it was to see my parents, the weekend did not go by without the topic of the Lavi making its way to the dinner table. My parents were constantly hearing about their evil son from acquaintances and family, and were more than a little distressed. My dad, who had never liked Arens, kept his own counsel. I knew that he did not agree with me on policy grounds, but he had always been scrupulous about not offering me free advice regarding any subject that came up in the course of my work.

My mom simply wanted all the bad news to go away. Like all Jewish (and non-Jewish) moms, she wanted her son to be seen only in the best of lights. I recall that during my years at the CBO, I once published a study that outlined the Soviet submarine threat in what was called the "Greenland-Iceland-UK gap." The study attracted headlines in Scandinavia and soon got *Pravda*'s attention: I had the pleasure of being accused of being a "jockey of imperialism, riding NATO's hobby horse."

The *Pravda* characterization was a fine turn of phrase, and, knowing my mother's special antipathy for the Soviet Communists, I phoned her with the news that I had been attacked in print.

"You were attacked?" she asked. "Are you all right?"

"I was attacked in print, Ma."

"Who doesn't like you? You didn't upset the navy, did you?"

"I was attacked by the Russians."

"Really? Where?"

"In *Pravda*, Ma."

"The Communists attacked *you*? In *Pravda*?"

"Yeah, Ma."

"What did they say?"

"They said I was a jockey of imperialism."

"Oh, wonderful. Did they mention you by name?"

"No, Ma, but everybody knows it was me they meant."

Pause. A long pause. "Oh." Another pause. "Well, when they attack you by name, let me know."

Mom's attitude had not changed much since then. She worried that the Jewish community would ostracize me. She worried about the effect it was having on my dad and my kids. She wanted me to back out—"Let somebody else do the job, you've done enough already" was pretty much the gist of her case. I listened, and carried on.

Back in Washington, we began to follow up on Bob Kasten's recommendation that we have more contact with the Hill. We scheduled a Lavi briefing for the Congress on July 16, just a few days before the rollout. We invited several members of the House Europe and Mideast Subcommittee, all staunch Lavi supporters: Gary Ackerman of New York, Bob Toricelli of New Jersey, Larry Smith of Florida, and Bob Dornan and Mel Levine of California (Lear Siegler, which built the flight control system for the Lavi—and for the F-16—was in Mel's congressional district). We also invited Charlie Wilson and Jack Kemp, and included, on our side, a number of the Lavi working group members.

We did not get all the members we had invited, but a number of them, including Jack Kemp, did show. Twelve of us were there to put the administration case. The presence of representatives of so many agencies was intended to dispel the notion—still circulated by the Israelis—that this was all a diabolical Pentagon plot, masterminded by a self-hating Jewish bureaucrat.

I devoted my remarks to an evaluation of the Lavi's financial prospects, asserting that the Israeli projection of $550 million in outlays was vastly optimistic, and that the real figure approached $1 billion. My colleagues then joined me in answering the members' technical questions. The questioning was not hostile, though several of the members, Jack Kemp included, made it perfectly clear that they intended to attend the rollout, and, in Kemp's case, to speak publicly in favor of the Lavi. Nevertheless, I came away with the sense that some of the members, possibly even including Jack Kemp, would not make a major fuss one way or the other over the Lavi's fate, but would vocally support the program while it was before the Congress. Later meetings with members of the House in particular convinced me that with the exceptions of a few diehard supporters like Larry Smith, Bob Toricelli, and Mel Levine, many members of Congress had no strong feelings about the program, and some would have preferred to oppose it. That few actually did so was, as the less sympathetic members told me, a product more of their desire not to alienate their Jewish voters and supporters than of any positive feelings they had toward the project.

Our briefing to the congressmen changed no one's mind. We did, however, succeed in altering the public impression that our effort was a Pentagon vendetta. As the *Jerusalem Post*'s international edition, which is widely read in the American Jewish community, stated, it was clear from our presentation that I was no longer merely the Pentagon representative, but spoke for the entire administration.

Some members of Congress actually did speak out against the plane. One notable example was Gary Hart, senator from Colorado and putative presidential candidate. Hart went so far as to suggest in a television interview that the Lavi was not necessary for Israel and that the country would do better to acquire American jets instead. Hart, a member of the Armed Services Committee, had a reputation as a champion of Israel's cause. His statement came as a shock and surprise to Lavi advocates, who identified all opponents of the plane with Israel's enemies. Hart's statement was even more irritating because he had only recently returned from Israel, where he had toured IAI and seen the Lavi, and because it was made just before the rollout. The Israeli response was churlish, reflecting an attitude that had permeated too large a portion of the upper reaches of the governing establishment: "The dogs are trying to bark," said an official who was too much of a coward to allow his name to appear in print beside the quotation.

Meanwhile, the issue of the blocked contracts finally came to a head. George Shultz had never agreed with Weinberger's position, and was upset that he had moved unilaterally. Rabin was even more upset. On the Friday before the rollout he went public on the issue of the contracts and said that Pickering and I had agreed to a "study for contracts" arrangement. He was furious that we had not delivered on what he considered to be our part of the bargain. "As long as there is no understanding with the U.S. on the release of those funds meant for the Lavi project which it is withholding," he asserted, "we will not cooperate in examining the alternatives proposed by the Americans."[3]

Rabin's outburst was accompanied by a formal representation by chargé d'affaires Eli Rubenstein to the State Department. The Israelis wanted their contracts approved and would not cooperate at all on the alternatives study until the logjam was broken. As if that weren't enough, at virtually the same time that Rabin was making his complaint public, Mel Levine and House Armed Services Committee chairman Les Aspin sent a letter to Weinberger protesting that "we have learned to our surprise that five contracts for Israel's Lavi project, with a total value of $70 million, are currently being held up." The congressmen also complained that a link was being made "between the hold on these contracts and the U.S. urging of Israel to consider alternatives to the Lavi fighter." They therefore advised DoD—in language that could have been written by the Histadrut Labor Federation—that "the U.S. suggestions should address not only the specific military and economic comparisons between the Lavi and possible alternatives, but also what may be termed the 'national' aspects of the Lavi project for Israel, including providing technological advances for its industry at large, nurturing a sophisticated work force, and preventing a brain drain."[4]

The onslaught from both Israel and the Congress prompted some frenzied activity at both State and DoD. State responded to the developments by writing a memo to NSC chief John Poindexter, requesting that the White House order DoD to approve the Lavi contracts. Navy Captain Jim Stark, who was then on the NSC staff, then passed me a copy of a letter to Cap Weinberger that had been drafted for the president's signature, which would order the release of the contracts. I duly gave the draft letter to the secretary's office.

Weinberger seemed to be holding firm, however. Concerned that my study would go nowhere if we continued to block the contracts, I enlisted Rich Armitage's support. We then made our case to the secretary, arguing that we had a much bigger fight ahead of us than the one over contracts. Recognizing that the issue was virtually settled anyway, and accepting Rich's judgment, as he invariably did, Weinberger caved at last.

In the meantime, we had fashioned a statement in response to the congressional letter. We argued, somewhat disingenuously, that we had never put a hold on the contracts, but were studying them more carefully. We also denied that we were urging Israel to consider alternatives to the Lavi. That was true only because we had not yet completed our study, however. We knew what was coming, and so did Congress and the Israelis. The letter was just one of the early shots across our bow. The heavy guns had yet to be fired.

Weinberger had now effectively defused the contracts issue, but the question arose of how to inform the Israelis. This prompted a new mini-flap with State. State's Near East Bureau proposed that Shultz send a "cover letter" to Peres advising him of our decision. In so doing, Shultz would have scored points with the Israelis, while Weinberger would once again have appeared to Jerusalem as the villain of the piece. I therefore vigorously opposed the idea, suggesting instead a joint cable from the two secretaries. My proposal was duly accepted, but not before Rich Armitage exploded upon learning that the cable was about to be sent without Weinberger's having seen the text. In the meantime, State now demanded that, in addition to the cable, Pickering make an oral statement to Rabin. I agreed, but there remained one more round for me to negotiate between State on the one hand and Weinberger on the other before all sides agreed to the dispatch of the cable.

I had barely had time to breathe after the cable incident when I was told by Larry Icenogle that Moshe Arens, in the course of an Israeli television interview relating to the Lavi rollout, had branded me as "the number one enemy of the Lavi project in the United States." I was floored that I commanded such importance in his eyes. I was particularly irked by the remark after having spent the better part of a week working to carry out my assurance (and Tom Pickering's) to Rabin that the contracts would be released so as to ensure the study went forward. I then learned that the Israelis had been led to believe that other "pro-Israel" officials had worked the compromise; no doubt they erroneously concluded that I opposed it.

I felt no better when a reporter for *Defense News* phoned to get my response, as an Orthodox rabbi, to my characterization by Moshe Keret, the president of IAI (the prime beneficiary of the just-released contracts), in a series of four-letter words. I could only reply that my religious preferences were my own business, and that Keret's indelicacy was unseemly for a major Israeli industrial figure.

I later learned that Keret had spoken to an American audience on hand for the rollout of the Lavi. I received a variety of reports of what he actually said; the only differences among them were the exact nature of the epithets he used. To their credit, Jim Roche, now a senior official of Northrop-Grumman, and Edward Luttwak, the internationally acclaimed strategist, proved to be true friends by standing up to Keret—in front of the assemblage—and telling him that they, unlike him, knew me personally, and that his outburst reflected badly on him rather than on me. Both his remarks and those of Arens also indicated to me that they were desperately afraid of the alternatives study, and of the damage to their program that was sure to follow in its wake.

Although Keret's remarks irritated me, I found it interesting that Jim Roche, who represented the "hated" Northrop firm, was among those receiving his message. The Israeli propaganda barrage increasingly targeted General Dynamics as the chief American industrial opponent of the Lavi program. Someone cited as an "Israeli official" told *Defense Week* shortly before the rollout that "'we know for sure' that General Dynamics has met with Israeli officials to try to kill the new fighter."[5] Keret told the press that "if anyone could win [from the death of the Lavi] it's General Dynamics."[6]

We had no evidence that American industry was trying to lobby the Israelis. Although it was possible that some low-level representatives or agents might have spoken to friends in the Israeli political establishment, the strategy of lobbying the MoD made little sense. Very few Israeli military officers were prepared to advocate the death of the Lavi even in private; fewer still would have wanted to be seen consorting with American industrial reps known to be active opponents of the system.

Industry officials never tired of reminding us—as they insisted to the American and Israeli press—that while they would love to have Israel purchase more American systems, they had been careful to stay out of the Lavi debate. The defense firms had been reluctant to help us analyze the Lavi alternatives, much less attack the Lavi itself. This was especially true of General Dynamics, which did not want to sour relations with MoD if the Lavi program actually did collapse, since it was most likely to benefit from the Lavi's demise.

The rollout came off without a hitch. On Monday night, July 21, with brass bands playing before a crowd of some two thousand invited guests, including most of Israel's establishment and Congressmen Ackerman, Levine, Smith, Toricelli, and Wilson, IAI rolled one of its two Lavi prototypes out of their hangar. (The second prototype was being held back for the first test flight in September.) Jack

Kemp was the only member of Congress to address the assemblage. His role as ranking minority member of the House Appropriations Foreign Operations Subcommittee, which approved the American Lavi funding, and his status as a possible presidential candidate in 1988 made him a special hero at IAI. The fact that IAI had granted Moog the subcontract for Lavi actuators showed that the IAI leadership appreciated the support they received; Moog is a Buffalo-based firm located in what was Kemp's home district.

The Israelis gave Kemp the red-carpet treatment. He met with Rabin, Peres, and Keret. He was photographed in the Lavi cockpit. For his part, Kemp said all the "right" things that the Israeli press and public wanted to hear. He paid his "highest compliments for the technological and engineering expertise of all here who have labored to produce this amazing fighter aircraft. It is a beauty to behold." Like Jewish mothers when their precious little boy conducts a flawless Torah reading at his bar mitzvah, Arens, Keret, IAI vice president Bully Blumkine, and Eini must have soaked up Kemp's words, sat back in their seats, and "kvelled."

Rabin and Peres also spoke to the crowd, who enjoyed a typically bountiful Israeli reception afterward. Peres insisted that the country had "no alternative." Rabin told foreign correspondents that since Israel was not requesting American forces during a war, it was Israel's right to determine its own defense policy and programs.

But not everybody was prepared to join wholeheartedly in the festivities. Avraham Ben Shoshan, the navy commander, allowed himself to be quoted in the *Los Angeles Times* as saying that "you could pay for the navy's entire shipbuilding program just with the accounting errors on the Lavi."[7] The absence of Ezer Weitzmann, who had started the project, was duly noted. So too was that of anyone from the American administration.[8]

Five thousand miles away, and within twenty-four hours of the rollout, I learned to my chagrin that the agreement that we had reached with Defense Minister Rabin regarding the alternatives study had not yet been finalized. Rabin wanted more than vague assurances about the contracts. In turn, Weinberger wrote to Rabin that we would release the contracts in exchange for full cooperation and openness on the alternatives study. The Israeli press reported that we were also asking for secret Israeli subsystems specially developed for the Lavi, and described Weinberger's note as "harsh."[9] I quickly found myself embattled on several fronts.

State was pressing us to release the contracts immediately, even though we had had no formal response from Rabin. In fact, we were not really sure where Rabin stood on the study. After I had won State's and Defense's support for a four-month timetable for the study, Rabin made it clear that even if he accepted the study, he did not like the timetable at all. We were reluctant to release the contracts until the matter of the timetable was settled.

Rabin also kept stressing that the study had to make provision for the impact of any alternatives on Israel's defense industry. It didn't help that he had never seen our terms of reference, although I had passed them along to Ben-Joseph for transmission to Tel Aviv. Rabin was therefore unaware of the fact that the study plan expressly provided for the preservation of Israel's high-technology workforce. This misunderstanding only worsened the already highly charged atmosphere surrounding the study.

It did not help either that the Israelis were also reportedly telling members of Congress that we had lied about the release of the contracts. I learned that Danny Halperin, not satisfied with the agreement that we supposedly had worked out, was pushing Bob Kasten to tack pro-Lavi legislation onto the debt ceiling bill that was soon to be considered on the Senate floor. Such an action would have forced the release of contracts without the Israelis having to make any concession to us, thereby possibly dooming the alternatives study.

Rabin finally sent us a note on August 4, stating that he would support the alternatives study on condition that we not hamstring the Lavi if Israel preferred none of the alternatives. He added a further condition that in case an alternative was chosen, the United States would cover all liabilities resulting from the termination of Lavi-related contracts ("termination liabilities"). These provisos seemed reasonable enough. We also acceded to Rabin's desire to delay the study, by setting the starting date for the fourth-month effort to July 26. We now planned for a presentation to the secretaries of state and defense by November 26. The new date represented about a two-month delay from our original timetable, and we suspected that the Israelis had sought the extension to account for a potential slip in the schedule for the Lavi's maiden flight. In line with the delay for the final product, we also slipped the first draft of the briefing to November 3, 1986.

No date was set for the actual briefing to Rabin and other senior Israeli officials. Indeed, until mid-August we did not even have an

Israeli "point of contact" for the study, nor had we finalized the agreed study parameters with the Israelis. Still, Rabin's note was a step in the right direction. One outcome was that Senator Kasten dropped any plans he might have made for new legislation on the Lavi.

It quickly transpired, however, that the MoD had still more conditions in mind. Ambassador Pickering informed me the day after he received Rabin's note that he had been handed another note, this time from David Ivri, that expanded upon Rabin's reply. The MoD director general wanted us to fund other new Lavi contracts, as well as contracts that had already been started. I let Tom know that I would check with our lawyers. Certainly it seemed reasonable, at least to me, that we fund contracts that were begun during the course of the study.

Matters did not end there, however. Following Ivri's lead, the Israelis continued to press their case, piling demand upon demand. On August 8 we met with Ben-Joseph and Eini to discuss the release of the contracts. It was a blistering July morning when Ben-Joseph, Eini, and most of the Lavi Alternatives Steering Group met in my office. The room was packed, and some of those attending were nodding off. Eini soon woke them up.

Eini and Ben-Joseph issued a demand for written assurances that in the event the study was delayed we would continue to fund Lavi contracts beyond the currently planned four-month study period; that we would fund those contracts until Israel made a final decision on the Lavi's future; and that we would approve all new contracts and modifications of existing contracts that might be submitted in the next four months. Eini added that we pay termination liabilities on all Lavi-related contracts that Israel had entered into with *European* firms.

The requests were not merely rephrasing a single concern. The Israelis, ever wary of our intentions, and not above a little cleverness of their own, wanted to be certain that we would not simply take four months to process old contracts and not deal with any new ones. They also wanted to be sure that *all* contracts would be approved. Since they were aware that we considered some of the contracts to be premature relative to the status of work on the Lavi, they wanted to ram those through as part of their price for cooperation on the alternatives study.

This was all really too much. I told Eini that we were not prepared to issue him a blank check. Eini shot back, "Why not?" We were stunned. I could see jaws dropping around the room; nobody was catnapping now. I replied that verbal assurances were enough. Eini was not done yet. If verbal assurances were acceptable, he asked, why not a blank check?

I couldn't believe I was engaged in this conversation. I had never encountered such chutzpah before. I could only wonder what some of the other members of my team were thinking. Whether he realized it or not, Eini was reflecting everything that was wrong in the American-Israeli relationship.

After the meeting ended we orally informed Ben-Joseph that we accepted his request that a delay in the study would justify continuing to fund Lavi contracts. We could not accept that we should fund contracts until Israel made a final decision on the Lavi, since the Israelis controlled the project's timetable. Nor were we prepared to approve new contracts sight unseen. In fact, we discovered that the New York office could not even identify the contracts they had in mind; for us to have approved them would have been ludicrous.

We insisted that each contract had to be evaluated on its own merits. Nor did we particularly like the idea of having Weinberger provide written assurances on the contracts. We saw this as an Israeli attempt to gloat over their victory and felt that this was an unseemly way for an American ally to behave. Accordingly, we advised Tom Pickering in Tel Aviv that he should inform Rabin that we would give fair, reasonable, and timely treatment to each contract request, and that he direct the New York mission, which seemed to us to be the most appropriate Israeli point of contact, to proceed with the cooperative portion of the study. In that vein, DSAA did provide verbal approval for ongoing work by the Bet Shemesh engine facility. We also made it clear that we would approve all contracts necessary to the smooth progress of the Lavi program during the course of the four-month study.

In a sense this latest flap was a test of Rabin's goodwill. If, as I expected, he was sincere in his desire to work with us and to sink the Lavi if the alternatives did prove to be more cost-effective, then he would accept our oral assurances and proceed. If, on the other hand, he simply wanted to stick it to the Defense Department, as Keret, Eini, and Arens preferred, he would reply to us in the negative. Rabin did not reply in the negative. I breathed a sigh of relief, and we moved ahead on the alternatives study.

Rabin's decision to support the study attracted widespread speculation in Israel that he approved the study because he really opposed the Lavi.[10] It also prompted a renewed discussion of alternatives. One of the more striking Israeli commentaries was by a research fellow at Tel Aviv University who produced a chart that compared the costs of civilian and military projects. Heading the list was the Lavi, whose $15 billion cost was equivalent to the cost of completely overhauling and modernizing Israel's high-technology sector. The researcher noted that $6 billion could be saved by buying American aircraft under license: such savings exceeded Israel's expenditures on the Lebanon War and could provide housing, infrastructure, and jobs for 200,000 people, as well as the proposed Tel Aviv metro system.[11]

The major American papers and newsmagazines also were devoting ever increasing attention to the story. For example, the *Washington Post*'s unflattering page-one portrait of the Lavi program included a quotation from a former State Department official who likened the creation of the Lavi to the "story of the stone soup." In the old tale, a man offers his stone to cook soup for a gullible stranger. He then requests water, carrots, onions, meat, and seasoning. Soon the soup has become a beef stew at the stranger's expense. The former official said he was reminded of the story when the Israelis proposed the Lavi in 1981. "They were going to build this airplane . . . all they needed was American technology and American money."[12] That it was a former diplomat, rather than a DoD official, relating the tale made it even more compelling. It was becoming increasingly clear that most Americans who were involved with the project recognized it was nothing more than a rip-off.

The IDF leadership was also becoming more outspoken on the issue. The defense budget was undergoing considerable pressure, to the point where Chief of Staff Moshe Levy contended somewhat cryptically that even if the Lavi was killed, "this would only be a partial answer" to the defense establishment's problems. "A senior Air Force source" was less equivocal, saying that "if it is now already clear that even canceling the project will not solve the budgetary problem, we should cancel it that much sooner."[13]

Many of us assumed that the Israeli air force "source" was Avihu Bin Nun, the air force's chief planner. Bin Nun could not go public, however. Earlier in July, Chief of Staff Levy had made it clear to his unruly troops, including his deputy Dan Shomron, that public criticism of the Lavi no longer would be tolerated. The Lavi's opponents scurried for cover; privately, however, they continued to lobby for the cancellation of the program.

Given Bin Nun's feelings, which were shared by many, indeed

most, of the air force's officer corps, we were not surprised that the air force was perfectly willing to cooperate with us on the study, though the commander, General Lapidot, made it clear that he was only following Rabin's orders. Lapidot had always been an outspoken Lavi advocate, so his "following orders" line seemed natural enough. It also protected him against the wrath of the once and possibly future defense minister and Lavi advocate Moshe Arens.

I learned from Vern Lee of General Dynamics that the defense minister's economic adviser, Zvi Tropp, also saw little point to the study, but not because of his support for the Lavi. Zvi felt that the most likely alternative was an "off-the-shelf" F-16, that is, one that could be bought without any new Israeli participation. It was therefore a waste of time to study other alternatives. Partly because of Vern's report I decided that it was best not to include a purely American F-16 in our alternatives, at least at the outset. If we did so, it not only might appear to be covering ground that the Israelis were fully aware of, since they operated F-16s, but also would seem to be a hard sell on behalf of GD.

Although Rabin had now agreed to cooperate with us on the alternatives study, little was expected to happen until an Israeli team could interact with our air force analysts, as well as the contractors, at a meeting scheduled at Wright Patterson Base (the Aeronautical Systems Division's headquarters) on August 18. Still, the air force was getting very nervous; time was running out on the November deadline. The air force leadership had not moved quickly to organize for the study and had lost the better part of July, and I had privately become so despondent that I had contemplated canceling the entire effort. It was only on July 24 that the Lavi Alternatives Steering Group formally met with the ASD team to launch the four-month effort. Moreover, Systems Command, ASD's parent command, made it clear to us that if the Israelis delayed much after August 18, the study could not be completed by the end of November.

On August 13 I sent Abraham Ben-Joseph an extensive list of questions that ASD felt needed to be dealt with in the course of our new study (a copy was also sent to our defense attaché in Tel Aviv). The questions were exhaustive. They addressed issues such as the number of planes Israel required (we assumed it was still three hundred), major program milestone requirements, and definition of the plane's configuration. We had detailed questions about the Israelis' desired concept of operational support, addressing such matters as basing, runway types, and maintenance philosophy. We were concerned about the mission capabilities and profiles that the Israelis sought. These included matters such as weapons load, mission profile descriptions, performance, and survivability requirements.

We needed to have a full understanding of the state of development of the avionics systems in the Lavi and of avionics subsystems, and arrangements for support. We asked for information about reliability, maintainability, and price.

We hoped that the Israelis would be prepared to address these questions at the Wright Patterson meeting. We had other questions too. These included a request for a very detailed description of the Lavi, including drawings of the aircraft and its avionics installation, structural design criteria, armament, and electrical information. We needed to know more about the Lavi's mission data processing system, the control of its combat and weapons suite, and its management of stores and weapons. We recognized that the Israelis could not furnish all of the information in so short a time frame, and did not press for answers on these details relating to the Lavi. We also informed the Israelis that they need not develop a special format for providing the information. We cared more about the details, not the manner in which they were presented.

Ben-Joseph replied quickly enough, but said that in view of the questions we posed, August 18 was an impractical date for a meeting. The chief of the New York purchasing office offered to give me a better sense of timing on that day, and in effect requested that I sit tight until then.[14]

On the same day as I had written to Ben-Joseph, I issued detailed guidance to the air force regarding our approach to the alternatives study. I reiterated that the study's objective was to provide alternatives to the Lavi program based upon derivatives of U.S. aircraft configured to meet Israeli mission requirements. To make the alternatives financially attractive, I established a cost constraint of $475 million, that is, an annual average of $75 million below Rabin's cap. Within this cap (which was eased somewhat if expensive coproduction options were to be offered), the alternatives were to maximize Israeli components, while meeting mission requirements, yet minimize the risk of integrating those components. The air force was also instructed to include the cost of all special premiums, such as capital expenditure, within the cap. The intent was to ensure that the Lavi's supporters could not complain that we had excluded certain costs from the estimates to avoid the cap.

We indicated to the air force team that it should inform the contractors that while we could not hold them too tightly to their cost estimates, they did have an implied obligation to stand behind them. Only about one-fourth of those estimates really involved factors unknown to the contractors. We suggested that to protect themselves, contractors' estimates should be conservative on the cost of integration and similar unknown factors.

We issued a variety of other directives affecting the cost of the alternatives, though no specific guidelines regarding contractual relationships, which could be worked out between the contractors and the Israelis. The study was to be done in constant fiscal year 1984 dollars, to be comparable to Israeli Lavi estimates. We told contractors that costs were to be "the most accurate estimate you can generate in the time allocated." We wanted the contractors to include termination liability costs (per Rabin's concern); they should estimate the costs of both direct sales and sales through the U.S. government, and propose the cheaper approach. We also wanted General Dynamics to assume that, in accord with the terms of America's previous sale of F-16s to Belgium, Norway, Denmark, and the Netherlands (the European Program Group), 15 percent of the F-16 work would go to industry in those countries. Finally, we indicated that the contractors could propose as many options as they wished, as long as they met all the other guidelines we had set.

The Israelis finally came to Wright Patterson on August 25. Seven MoD officials, led by the director of engineering in the Lavi program office, met for three days with a team of both air force and contractor personnel under the leadership of ASD. There were difficulties from the outset: the Israelis had not been aware that contractor representatives would be attending. They argued that their data, which related to the Lavi, had been meant for U.S. government eyes only. They added that Lavi data were not necessarily required by the contractors for their study.

The more likely reason for the Israelis' reticence was that they did not want too many Lavi details to fall into the hands of potential competitors. Of course, that was what the alternatives exercise was all about. The U.S. side had always intended to share all its data with the contractors. Only through a thorough understanding of the Lavi's missions, capabilities, and technical characteristics could the contractors formulate viable alternatives for Israeli consideration.

In the end, the two sides reached a workable compromise. All contractors met with both the U.S. government and Israeli teams for introductions and general Lavi presentations by the Israeli team. The contractors met with the Israelis on an individual basis to obtain specific information that they required. On the other hand, the Israelis provided the contractors with no pricing data and gave their detailed responses to ASD's questions only to the government team.

Each of the contractors surfaced its particular concerns during the individual meetings. GD was at great pains to emphasize that it had only joined the alternatives study at the specific behest of the U.S. government; in other words, it had been given no choice. The GD team recognized that it could not meet all of the Lavi's operational requirements without building another Lavi. It therefore intended to offer an alternative that met the Lavi's *major* requirements. It also intended to offer the Israelis a significant amount of technical work on the F-16, possibly a "Peace Marble III," and requested both the greatest possible detail on avionics so as to formulate an attractive work package for the Israelis and meetings in Israel to establish the requisite business arrangements.

McDonnell Douglas also had a series of requests for data from the Israelis, relating not only to avionics but to runway types, shelters, and other support factors. The latter requests stemmed from McDonnell's decision to propose not only the F-18, but also the short-takeoff-and-landing AV-8B in both "generic" (i.e., basic) form and with options for Lavi avionics as well as production options in Israel. McDonnell also assumed that final assembly and test in Israel would be an attractive option.

Northrop asked if the F-20 would be a viable candidate; the Israelis would not provide a straight answer. Accordingly, Northrop decided to go ahead with an F-20 alternative, and, like the others, plied the Israelis with technical queries about avionics, support, and performance.

The Israelis provided the U.S. government team with a book of responses to the questions they had received. These were to be treated as confidential and not made available to contractors without Israeli consent. They established an official channel of communications with the contractors via their New York office and our office. The communications line ran from the Lavi office, through the New York mission, through my office, to air force headquarters, to Systems Command headquarters, to the study team, to the contractors.

In general, the meeting was a success. The Israelis had not provided all the information we had requested, notably details on avionics and cost, but at least had given the contractors enough material to get started on their work. The Israelis had also agreed to future meetings. I recognized the Israeli discomfort with the prospect of dealing with contractors, particularly as a group. This explained the cumbersome line of communication that had been agreed to, but we decided to work as quickly as we could within the constraints that the exchange had imposed on us. It was now up to us to ensure that the contractors provided to Rabin viable alternatives that would permit him to jettison the Lavi.

Two days after the conclusion of the Wright Patterson meeting, ASD headquarters circulated within the Pentagon an extremely detailed draft letter of request that outlined the manner in which the contractors were to frame their options. The draft letter also included the guidelines that I had issued to the air force on August 13. I circulated the draft to my steering group and sent a copy to Ben-Joseph. The letter was sent on September 4; the contractors therefore had about six weeks to prepare the complete version of their alternatives.

While work progressed on the alternatives, I became convinced that it would be worthwhile to include some additional "sweetener" in the study apart from whatever Israeli contributions could be identified for the various American airframes. I suggested to Fred Iklé that we offer the Israelis cooperation on anti–tactical ballistic missile (ATBM) defense, a mission that was getting increasing attention in DoD.

Although I had not been involved in Cap Weinberger's attempt to win allied cooperation for our Strategic Defense Initiative, I was cochairman of a Pentagon group that was looking at short-term ATBM programs. We knew the Israelis were interested in SDI; they faced a tactical ballistic missile threat from their neighbors. Israel had been among the first to respond to Weinberger's offer of cooperation. On the other hand, while the Israelis had expertise, they had no money to speak of. I had asked the SDI office to identify projects that we could work on jointly with the Israelis, with their share funded by our security assistance budget.

Iklé agreed to discuss the matter with the SDI office's director, General Jim Abrahamson, and also agreed that I should initially keep the ATBM discussion separate from the air force's examination of the alternative airframes. I also said that I would not raise the issue with Rabin during my next visit to Israel, which was scheduled for September, so as not to create the impression that we had funding readily available for the MoD's use. Again, Iklé concurred.

Even as we had finally begun to make progress on the alternatives study, we learned that the GAO was likely to issue a report that would give something to both sides in the Lavi cost dispute. Derek Vander Schaaf, the DoD deputy inspector general, whom I had known since my CBO days when he was on the House Appropriations Committee staff, sent a memorandum to SecDef reporting on "informal discussions" with the GAO investigators. The GAO team had just returned from their three-week study trip to Israel. They had managed to attend the Lavi rollout, as well as discuss the program with both MoD and industry representatives. They returned home concluding that the Israeli estimate was probably more accurate than ours, since our postulated labor rates were too high. On the other hand, and, in Derek's view, of greater importance, "the GAO has apparently also concluded that Israel will have a major cash flow problem if it continues to pursue the Lavi program as it is currently structured . . . even assuming the lower cost estimate . . . the Lavi program will cost Israel about $1 billion per year from 1988/89 through the year 2000. The GAO claims that the Israel economy cannot support such a drain." Derek went on to say that the GAO apparently was going to take the position that "during these times of budget constraints and high deficits, it is neither prudent nor realistic for Israel to pursue (or the United States to support) a program that will have to depend on major increases in U.S. assistance."[15]

I could only hope that the GAO followed through on what it had told Vander Schaaf. I knew that he had a good relationship with the agency, and that he was a straight shooter. He was reporting exactly what he had been told. I also knew that he thought the Lavi was a big waste of money, and that he certainly would do nothing to dissuade the GAO from publishing its preliminary findings.

The Israelis were already generating stories that the GAO was vindicating their position, and that our estimates were all wrong. No doubt every ounce of pressure would be exerted on the analysts to get them to drop their key conclusion: that the Lavi was unaffordable. Derek had noted that the GAO was going to issue a report on the cost issue in September. I was certain the Israelis, through their friends on the Hill, would seek a delay in the report so as to have the conclusions reflect a position more favorable to the Lavi.

14

The Navy Study
Is Completed

During the first week of September I held a "nonmeeting" with Prime Minister Shimon Peres's chief economic adviser, Amnon Neubach. Neubach felt that the end was in sight. He urged me to press ahead and told me that Peres and Rabin had agreed that the United States had to provide the excuse for the program to be terminated. I doubted his story. Peres and Rabin rarely agreed about anything. Now they seemingly were in agreement that I should run interference for them. Somehow it didn't add up.

Two days after I met with Neubach, the Lavi Alternatives Steering Group met to review the contractors' progress. Vern Lee of General Dynamics opened the meeting with a presentation of the differences between the 1982 GD proposal and that which the company was now going to put forward. The key to the new proposal was GD's willingness to let the Israelis work on the complex, and costly, forward fuselage, where the "black boxes" would be housed. Lee also raised the European Program Group production issue, which threatened to complicate GD's ability to formulate a viable Israeli work share for its alternatives. We agreed to reverse my August 13 directive, and GD was permitted to exclude EPG participation. State offered to look into the possibility of exempting the Israeli F-16 from EPG commitments because of the changes to the aircraft. We all appreciated this display of diplomatic craft (and craftiness).

We continued to hear from the contractors as they geared up for the trip to Israel, which we had agreed should take place not later than September 21, if the study timetable was to be met. Each contractor sent us a list of desired meetings, subjects, and personnel who would be going on the trip. We also had requests from the contractors for clarification about restrictions on technology transfer: under the terms of current legislation, it was forbidden even to *discuss* any sort of technology-sharing arrangements, much less technical data, without a special exemption from the State Department. Upon consultation with State and DSAA, we replied that clearance would be given to discuss coproduction and licensed production, aircraft capabilities, and interfaces for the employment of required weapons.

Although the contractors felt that they were under the gun, it had become widely known that the first flight of the Lavi was being delayed at least a month, because of Lear Siegler's problems in delivering the flight control computer. I doubted, therefore, that the Israelis would let us present the study before the first flight. That would give the contractors a bit more time to complete their portion of the work. Nevertheless, the Israelis had said nothing to us about delaying the timetable for the study, and for the time being, the November deadline held firm.

While nothing was said about delaying the study, delaying the visit was quite another matter. The fact that the steering group approved the visit date for the contractors meant very little. The Israelis had to agree as well. And the Israelis, despite Rabin's commitment to the study, were still not in an agreeable mood.

The problem was Menachem Eini. I had written to Ben-Joseph in New York just after the steering group meeting requesting the contractors visit from September 21 through 28, with each contractor allotted two or three days for individual meetings with MoD and industry. In an effort to display our own openness, I also invited him to our future steering group meetings. It was Eini who replied. Ever up to his old tricks, the Israeli phoned me on September 17, only four days before the contractors were scheduled to depart, to inform me that there were difficulties with the trip. He claimed that the schedule was too compressed to enable the Israeli air force to organize the Israeli side. I found his assertion a bit surprising, since we assumed that the Israelis had all of the details we needed, and we had given them nearly three weeks' notice, almost half the time we had to complete our entire effort! Eini finally agreed to slip the trip by only four days—he must have needed to save face, I suppose.

The second difficulty was a real one, and very annoying at that. Eini said that IAI president Moshe Keret was unwilling to cooperate with the United States on a variety of matters, including IAI's preferred contractual arrangements in the event the F-16 was selected. "IAI views continuation of the Lavi as a life-and-death situation," Eini said.

It was astonishing that the president of IAI, a government-owned company, was ready to defy the minister of defense. I was not going to let Eini get away with this one. I told Eini that I preferred not to have Caspar Weinberger contact Minister Rabin to force IAI's cooperation. Eini backed off: "of course Minister Rabin could force IAI to be more forthcoming," but he hoped that "such a situation could be avoided." Eini agreed that IAI would be forthcoming on all information relating to systems and subsystems that the contractors required for their analysis. As a conciliatory gesture—which Eini did not deserve—I agreed that if possible we would avoid the issue of IAI's preferred solution.

Although IAI was unable to short-circuit our study, it was also taking other steps to undermine the rationale for our effort. On August 31, 1986, after some negotiation, IAI signed a memorandum of understanding (MOU) with Grumman regarding future business arrangements for marketing and producing the Lavi outside Israel. The agreement provided the groundwork for potential coproduction of the Lavi in the United States and for Grumman to assist the Israelis in marketing the plane worldwide, including in the United States.

The Lavi's supporters, unable to rid themselves of the illusion that the plane would have an export market and even could be sold to the U.S. Air Force, hailed the agreement as a major step forward. The promise of an export market would justify Israel's lower cost estimates, if only because of potential economies of scale. The case for an alternative would thereby be seriously weakened.

Grumman itself was less ambitious about the MOU, however. A senior company official wrote to me that the MOU provided the framework by which Grumman had agreed only to *study* the feasibility of a possible business arrangement for production of the Lavi for requirements external to Israel. The Long Island company clearly recognized that the Pentagon would not go out of its way to support any arrangement for marketing the Lavi, and certainly would be niggardly about granting the requisite licenses without which technical cooperation, much less exports, would be impossible.

Whatever the Israelis may have thought about their prospects with Grumman, we saw no reason to change our strategy, or our timetable. The Lavi's own timetable was another matter, however. By the time Grumman's competitors went off to Israel on September 25, 1986, there were new reports about a postponement of the Lavi's first flight. Lear Siegler was promising delivery of the flight control computer "within the next two weeks." After the computer's delivery there would be at least a month of ground testing before the computer could be inserted onboard the plane. It looked as if the plane would not fly before December at the earliest.

I had felt for some time that the flight control system would prove to be the Israelis' Achilles' heel. Lear Siegler was also producing a system for the Swedish Gripen, and it was amusing to hear Eini and his Swedish counterpart, the Gripen program manager, each insist that his program would not encounter the difficulties that his counterpart faced. In reality, the Gripen program was delayed several years as well; the plane began to enter the Swedish air force structure only in 1994.

The contractors' meetings in Israel went off relatively smoothly; Israelis are terrific at cramming in the maximum number of meetings in the minimum time available. No business discussions were held, and IAI must have breathed a sigh of relief. The Israelis also offered little by way of avionics development costs or schedules, despite my supposed agreement with Eini. On the other hand, the contractors were able to survey Israeli industry in general terms, and received answers to their technical questions from the Lavi program office. It was not all we had wanted, but enough to allow the contractors to maintain the momentum of their effort. Under the circumstances, that was the most we could have hoped for.

There were often times during the summer of 1986 when I felt as if I were totally consumed by the Lavi. The unceasing controversy had literally gotten under my skin: I discovered in late summer that I had developed a mild case of psoriasis, which, my doctor asserted, often was a function of stress. Was I working on anything particularly stressful at the moment? I could only laugh. Psoriasis comes more easily than it goes; the controversy may be over, but my skin is no better.

The month of September brought some respite from the fighter program, however. Shortly after our Lavi Alternatives Steering Group meeting on the fifth, I went off to Tokyo for a negotiating session on the installation of a new American system in Japan. My interaction with the Japanese contrasted sharply with my dealings with Israelis. Japanese officials are unfailingly polite; chutzpah is not in their lexicon. Nor do they argue with each other as Israelis are wont to do, even before non-Israelis. The facade, and often the reality, of consensus is critical to Japanese dealings with foreigners. They will not move on a given issue until they are certain that everyone with a stake in that issue has been accounted for and has agreed to the next step to be taken. This approach is operationally time-consuming. So too is their officials' tactic always to require consecutive translation in any discussion, even if, as is invariably the case, they understand English perfectly well. Such delays tend to wear down American negotiators; the history of our failed official dealings with the Japanese speaks for itself.

Americans also tend to assume that once agreement in principle is reached, the details can be quickly worked out. That is not at all the case with Japan. My task was to obtain agreement in principle for the installation. It subsequently took the United States several years to reach a detailed agreement to carry it out. The record of U.S.–Japanese negotiations is clear on this point as well.

Prime Minister Shimon Peres came to Washington just after I returned from Japan. His round of visits included the Pentagon; we listed the Lavi among the secretary's talking points. The discussion changed neither man's opinion.

I chatted briefly with Peres at a State Department dinner that Vice President and Mrs. Bush hosted. Peres asked me in Hebrew how I was, and then, flashing his warmest smile, "And how is the Lavi?" I responded that I was fine, as was the Lavi. Peres smiled again, and there the matter lay.

It was also during September that we completed a draft of the naval modernization study. I circulated the document on September 22, 1986, requesting replies in about two weeks, and scheduling dissemination of the final report ten days after that. I was pleased with the way our navy and the Israeli navy had cooperated both with each other and with OSD on the study. I hoped it would be well received in Israel.

It was only in late September that Thyssen, HDW's primary German competitor for the submarine portion of the program, finally briefed our team on the yard, its capabilities, and its view of the Dolphin program. Despite the delay, we took Thyssen's interest as a sign that the German government was not going to stand in the way of the program, if it was managed discreetly as a subcontract let by an American shipyard.

We learned at about the same time, however, that Bath Iron Works had dropped out of the competition for the surface ship modernization program. In a letter dated September 25, 1986, the company's president complained bitterly about Israeli behavior. "We are at best confused by the environment . . . our expertise is engineering and shipbuilding—not finance and contract intrigue."[1] The company complained that the Israelis wanted Bath to finance the first two years of the modernization program, as well as provide offsets for the cost of the submarines. In addition, Bath was told that "it would be very desirable" if it took an equity position in Israel Shipyards, which was slated to build the subs.

I was not surprised by the Israelis' behavior. We had suggested an offset arrangement, which involved work to be undertaken by Israel to be credited by Bath Iron Works against the cost of the submarine construction. The offset was to be a form of barter—Israeli goods and services in exchange for those provided by the American shipyard. Of course, what the goods and services might be and to what extent Bath needed those offsets remained unspecified. In many instances, American defense companies turned to specialized firms which undertook to market foreign products as varied as wine or perfume in exchange for defense sales to countries producing those goods. The Israelis tended to prefer either the purchase of their domestically produced defense goods and services, often as subcontractors to the American producer, or better yet, investment in their research and development programs.

In this case, however, the Israelis had gone several steps further—they wanted financing, *in addition to* barter arrangements. After all, why not go for a leg if an arm was already available? Bath was not used to these sorts of dealings. American shipbuilders are quite insular, having the navy as their primary, and sometimes sole, market. The other candidates took heed of Bath's experience, however. In the end, at least to my knowledge, Israel Shipyards received no financing from any American source in connection with the modernization program.

Normally, September was also High Holiday month, which often meant seven days away from the office and early departure on the eve of each of the holidays. This time, however, because of the nature of the Jewish calendar, which added a month to the previous year, the High Holidays fell in October. Because the Jewish calendar is organized around both lunar months (like Islam) and a solar year (like Christianity), Jews add a thirteenth month every second or third year so that the lunar and solar elements of the calendar will remain synchronized. The system, once based on sighting the new moon, was set in concrete by the Talmudic scholar Hillel; its intent was to ensure that Passover always fell after the spring equinox. Otherwise, Passover could fall at any time of the year, as do Ramadan and the other holy periods of the Islamic calendar.

The extra month usually pushes the Jewish New Year to late September; this year it was even later. And it was a pleasure to be away from the office. Even the long services—the longest of the year—were a welcome respite from the hyperactive pace as we prepared to issue the submarine report, crunched ahead on the alternatives study, and addressed the many other issues, notably the defense planning process, that were the daily workfare of our organization.

As the naval modernization study neared completion, we got word early in October that, despite having issued requests for proposals for design and construction of both the corvettes and submarines, the Israelis were once again assigning higher priority to the submarine portion of the program. This ran counter both to our evaluation that there was not as urgent a requirement for the Dolphin submarines as the Israelis alleged and to our economic interest in having the Saar boats built in the United States. Accordingly, I reiterated our position to the new Israeli defense attaché, Amos Yaron, and the naval attaché, Chaim Shaked. I noted that we had agreed with MoD that the modernization program was a package deal comprising both those ships and the submarines. I added that if the Israelis continued to support the package deal that the United States preferred, DoD was prepared to assist them in finding ways to finance even those elements of the program that were to be produced in third countries.

I found myself preaching to the choir. Shaked wanted the entire program to go forward; the pressure for the submarines appeared to be emanating from unnamed elements within the MoD. Yaron's support for our position was somewhat more surprising, but Yaron was basically a straight shooter who evaluated each issue on its merits. Amos looked the part of a grizzled veteran, with a craggy face and curly black hair. He had been implicated in the 1982 Sabra and

Shatila massacres of Palestinians, and his appointment as attaché to the United States and Canada had sparked some controversy in both Ottawa and Washington. I never became involved in the dispute over his background, and I found him to be both generous with advice and an invaluable conduit to Rabin. He also became a powerful ally. Amos made no bones about the Lavi; he didn't like it and wouldn't support it.

We completed the naval modernization study on October 31, 1986, though formally it was not done until we had first briefed it in Israel. Like the Lavi report, it was lengthy and highly detailed, running to 138 pages plus six appendices. The executive summary was unclassified, just as that of the Lavi study had been, so as to permit the widest possible distribution.

To underscore the contrast with the rancor surrounding the Lavi debate, the summary emphasized the teamwork that had gone into its preparation, both between the Israelis and the American team and among the various U.S. government agencies that constituted that team. It stated that the Israeli program for four Saar corvettes and three Dolphin submarines was extremely ambitious in terms both of the capabilities it sought to add to the Israeli fleet and of the cost of the systems that it proposed to acquire. It reiterated the American view that the need to replace aging Saar III boats with the Saar Vs was much more urgent than the need to replace the Gal submarines, although it validated the requirement for the Dolphins.

The report stated that in the event that Israel initiated the Saar/Dolphin program, it should make every effort to minimize European content while maximizing that of the United States. It recommended that Israel combine the two programs as a package, with a single U.S. prime contractor who would subcontract for German construction of the first Dolphin as well as for hull sections of the second and third submarines. The second and third submarines could then be assembled in Israel.

The report recognized that Israel had never built a submarine before. Nevertheless, it noted that both the German shipbuilders, HDW and Thyssen, had considerable experience in developing leader-follower programs with states that had no shipbuilding experience. Some of these states were in the Third World, with less technical expertise and ingenuity than Israel.

Finally, with respect to funding, the report indicated that there was only a $100 million gap between the revised Israeli estimate of $1.25 billion and the higher U.S. estimate. This was in sharp contrast, of course, to the dispute over the Lavi cost estimates. The report also stipulated that Israel could not expect increases in security assistance to support naval modernization, nor would it get special dispensation for financing European elements of the program. Naval modernization would have to compete with other programs (the Lavi was not mentioned but obviously sprang to mind) for funding support. On the other hand, the U.S. and Israeli teams agreed that American prime contractors should be required to generate indirect offsets to the value of 75 percent of all work that would be undertaken in Europe, thereby easing the financial impact that such activity would have upon the Israeli balance of payments. Such offsets would ideally be organized so that Israel could provide goods and services to Germany in direct payment for the work done there.

The report noted that cash flow would be a problem throughout the lifetime of the program. The Israelis would face onerous annual operating costs for both the ships and submarines; we estimated that the Saar V maintenance costs would be double those of the corvettes they replaced, while fuel costs would quadruple. There was also considerable uncertainty about other elements of the program's annual costs and about its overall life cycle costs. The report consequently recommended delaying the Dolphin program by two years to ease the cash flow burden. This was a blow to those Israelis, especially within the MoD, whose higher priority was the submarine program rather than the corvettes. We were soon to learn just how much they resented this recommendation, as well as our suggestions regarding financing, when we visited Israel. Still, the overall message of the report was a positive one: we had validated the Israeli requirements for both classes of ships.

We briefed the study to senior DoD officials, including Cap Weinberger and John Lehman, during the first week of November 1986; reaction was positive. We therefore returned to Israel to brief the study to the MoD and Israeli navy leadership. Just prior to our departure on November 8 we also presented a summary of the report to Abraham Ben-Joseph's deputy at the New York purchasing office. The deputy, a veteran MoD bureaucrat on the verge of retirement, was noncommittal, which was an improvement over Israeli reactions to our previous Lavi study.

We had good news and bad news for our Israeli interlocutors. The good news was that for once there was no conflict over cost estimates. Ours were simply too close to those of the Israeli navy, which had revised its own estimates upward based on our analysis. The bad news was that we were opposed to the simultaneous commencement of the submarine and corvette programs; nor were we going to make any pronouncement regarding American financial support for Israeli naval work in Europe.

I also brought news of which the MoD already was informally aware. The U.S. government remained firm in its determination to terminate the program of "directed offsets," whereby American industry was compelled to buy Israeli goods and services whether or not they were actually needed. Even our own proposal for offsets as they related to the naval modernization was formulated in a manner quite different from long-standing "directed offsets," since it involved Israel's possible provision of goods and services to German firms in exchange for the cost of submarine construction.

As we had expected, the navy leadership was generally pleased with the report, which we briefed to them on November 10, our first business day in Israel. Avraham Ben Shoshan, the navy commander, and his colleagues recognized that the study strengthened their hand vis-à-vis the MoD leadership, whose enthusiasm for the modernization program was rather lukewarm.

Matters were not as smooth when it came to financing the subs, an issue which fell into the bailiwick of the MoD staff. After our meeting with the naval staff, we walked the few hundred yards to the main building, where we met with David Ivri and his chief official for naval programs, Avraham Oren. Ivri was noncommittal. His eyes did not light up over submarines and corvettes as they did over mention of the Lavi.

Oren did not like our suggestion that we delay submarine construction for two years to ease Israel's cash flow problems. He also dismissed our suggestions that Israel could finance the boats by selling goods and services to American or German prime contractors. It didn't help that we were advocating this approach even as we sought to roll back the directed offsets program. The Israeli government was looking for direct cash payments, not a convoluted stream of funding that eventually would wind up in the hands of the Germans.

I found it ironic that the Israelis, of all people, were suddenly insisting that money was not fungible. It was widely known in Washington that Israel was able to juggle its internal accounts on the basis of its expected dollar revenues from security assistance. Monies that would otherwise have been spent on defense were constantly shifted to the "shekel budget" because defense activities were covered in the security-assistance-based "dollar budget." In effect, U.S. dollars were covering every aspect of Israeli fiscal activity—and they still do. Yet when we raised a similar idea, namely that funds received for Israeli goods and services be diverted to cover submarine-related expenditures, Oren said the idea was unworkable.

Always eager to stir the pot, the press picked up on the differences between ourselves and the Israelis over the submarine construction schedule and the financing for the project.[2] An anonymous "defense source" dismissed our recommendations and stated that he foresaw "long and arduous negotiations." He added that the real impact of our suggestions was to force a trade-off between the naval project and the Lavi, or alternatively, if the Lavi were to go ahead intact, to delay the modernization program until 1989, when some foreign military sales funding might be available.[3] The source said that Israel was disappointed because it had hoped to employ FMS funds to move ahead with both projects simultaneously, though how it could have done so without an increase in FMS funding—which was not in the cards—was beyond my comprehension.

In fact, the differences between us were by no means as deep as reported, or as some MoD officials might have hoped. The Israeli navy, which in any event had no sympathy for the Lavi, was aware of my commitment to Yaron and Shaked. In private discussions with Ben Shoshan and his deputy, Michah Ram, I renewed my promise to the attachés that I would do whatever I could to get Weinberger to authorize some third-country FMS funding for the submarine program. Perhaps the fact that my promise had not leaked to the press constituted the best evidence of the closeness of our working relationship with the Israeli navy.

Upon returning to Jerusalem after my meetings at the defense ministry, I had an exceedingly pleasant visit with Shamir at his offices near the Knesset. He was now prime minister, and enjoying every minute of it. He did not stop smiling the entire time I saw him. I had the impression that he felt that he had outfoxed Shimon Peres, the leader of the Labor Party, with whom he had switched places. In accordance with the agreement that governed the "grand coalition" between the two major parties that constituted Israel's National Unity government, Peres was now foreign minister. It seemed to me that Shamir would be in charge for quite some time.

Shamir was very complimentary of our naval modernization study. He said that he supported the plan, and added that he hoped that the defense establishment would accept its recommendations. In practice, the decision was not really Shamir's. Rabin functioned as a proconsul in the defense ministry; his independence had been ratified in the original agreement for a National Unity government that Shamir and Peres had negotiated. Only when Rabin took the matter to the cabinet, as had already been the case with respect to the Lavi, could Shamir be expected to weigh in. Nevertheless, the prime minister's support was very important, and the fact that his support was reported in the press was even more useful.[4]

Only two people attended our meeting, Avi Pazner and Eli Rubenstein. Pazner, Shamir's press spokesman, is an extremely amiable and able fellow who has since moved on to be ambassador to France. No doubt he was responsible for briefings to the press. Eli, freshly arrived from Washington and now cabinet secretary, seemed a bit unsure of himself with Shamir. I saw no reason why that should be the case. At one point I told Shamir how wonderful Eli had been in Washington. Shamir, who has a very dry sense of humor, responded by saying, "Maybe we should send him back." Eli visibly blanched. Eli is extremely bright and, perhaps even more important to Shamir, is of Lithuanian origin. I knew that they would hit it off; which they did.

The next day I briefed Rabin, Ivri, Chief of Staff Moshe Levy, and senior MoD and navy people in Rabin's conference room. In contrast to my meeting with Shamir, I found Rabin exceedingly stiff, even by his own standards. He ignored my attempts at some mild humor about the report—I said it was now in "blue and white"—and generally displayed far less interest in the naval report than I had expected. Ivri helped turn the discussion once again to the issue of financing. Rabin emphasized Israel's budgetary constraints; I responded by citing our own financial problems.

Rabin and I reviewed the financing issue; there were no new ideas put forward by either side. I did note the report's suggestion that Israel build the last two submarines in Israel, but Rabin showed no great enthusiasm for the idea. Although I was by now used to Rabin's demeanor, I felt that he might have reacted more warmly to the naval report. Rabin was considered something of a savior in the Israeli navy; many years earlier, he had approved funding for the Gabriel surface-to-surface missile when he was under pressure to cancel it. Shimon Hefetz, his military aide, had sat in on the meeting; I asked Shimon afterward what was wrong. He could only reply that Rabin had other, more important headaches at this time. Iran-contra was breaking, with the Israelis linked to arms sales to the Iranians. In addition, Israel had become embroiled in a controversy with Britain over the Israeli kidnapping of Michael Vanunu, the former official who leaked details of Israel's nuclear program. Hefetz said things would be different when I came back to report on the Lavi alternatives. On that point I had no doubt he was right.

Rabin and I did not get into detail about the Lavi alternatives report, except to discuss in general terms when I expected to be back. The Lavi first flight appeared likely to be delayed again, and there was no chance of our presenting alternatives to Defense Minister Rabin before the plane flew.

"Defense sources" were quick to report their version of the meeting to the press, as they had the day before. Again, the emphasis was on our differences with the Israelis. The unnamed sources questioned the idea that Israel build its own submarines and emphasized that the decision to build the subs was not yet final.[5] It was becoming an Israeli refrain.

I decided that it would be best to clear up matters with a press conference of my own. Ambassador Tom Pickering agreed, and we organized a meeting for defense journalists at the American embassy. I walked the reporters through the modernization briefing, emphasizing all the while that our work had been done in concert with the Israeli navy and with the cooperation of the ministry of defense. Responding to the many backgrounders that had been provided to them by MoD officials, the reporters plied me with questions about the issue of financing. Several of them attempted to demonstrate that as long as the United States would not commit to financing German construction, we did not seriously support the modernization program. I pointed out that the discussion was really about a relatively small part of the overall $1.3 billion corvette and submarine program. The three submarines would cost about $150 million each, and at most 40 percent of that cost, or about $180 million, was at issue. Thus, the cost of European construction would amount to less than 15 percent of the entire program, and would be spread over ten years, which made it quite manageable. Moreover, if Israel built the second and third subs, the financial challenge would be even smaller.

The reporters were not convinced. Many stressed the financial competition with the Lavi; in fact, they saw my involvement in the costing of the naval program as just another aspect of my purported vendetta against the Lavi. I refused to comment on the plane during the conference or in individual interviews with reporters, other than to disparage the notion of competition between the programs, since the Lavi's costs swamped those of naval modernization. My reticence on the subject notwithstanding, the Lavi appeared in virtually every story relating to my trip. Perhaps the most noteworthy piece was an item in *Ha'aretz,* which detailed the despair at the Bet Shemesh engine plant in response to the decision to place the plant, which was supposed to build the Lavi's engines, into receivership. The story carried a front-page picture of workers demonstrating against what was widely viewed as the first step to the plant's closure. It recounted the plaintive recollections of the workers that Arens and virtually every other minister had promised them, had *assured* them, that the Lavi engine project would not die.[6]

It was wrenching to read about these workers. They had families to feed, clothe, and house. Some of them were the best and brightest of Israel's engineers; others were the Israeli equivalent of blue-collar workers. Many of them had no real idea of the politics behind the plane. None of these people was responsible for the mismanagement of the plant, nor for the financial machinations that led the state comptroller to open an investigation of bribery at the plant. All these people knew was what Arens and his henchmen told them: there was a bad American named Zakheim who was trying to take their jobs away. I felt for those people—who wouldn't?—but that was why we were constructing alternatives to keep them at work. To keep the Lavi alive was to invite the very financial and personal disaster that these people hoped to avert.

Among the many journalists who besieged me for information about the Lavi alternatives study was a young reporter who wrote for *Davar.* She told me that we were distant relatives; her mother had been born in Rozhenoi, my dad's town, and was a Zakheim. The woman had the high cheekbones that I could identify in all the pictures of the women in my father's family—none of whom I had met, since they perished in the Holocaust. I decided to look her mother up; she was an officer with Tel Aviv's Museum of the Diaspora, which I had always meant to visit. The museum is dedicated to the history of Diaspora Jewry, and is truly unique. The reporter's mother was extremely hospitable and arranged a wonderful tour for me. We were able to establish that she was something like my sixth cousin, several times removed. Perhaps we weren't "kissin' cousins," but we developed a pleasant friendship. As for her daughter, she maintained a purely professional relationship with me, which was too bad for me. Had she thought of me as a relative, perhaps she wouldn't have been as merciless when she questioned me about our analyses.

My hostess at the Diaspora Museum was not my only "long-lost relative" in Israel. I received numerous cards from people claiming to be related to me in some way. Since all Zakheims, Sackheims, Zackheims, and I think Zakims trace their ancestry to a Jewish martyr of the seventeenth century, they were relatives of a sort. I'm sure they all meant well, and no doubt were all decent people, but I couldn't help but wonder whether they would have been so eager to claim the relationship if I had simply been a visiting American tourist. In the event, I did not have the time to contact them, with one exception. One day I received a call from the speaker of the Knesset, Dov Zakin. Zakin was a leading left-winger in the Labor coalition who had long been associated with Mapam, the party to the left of Rabin's Mapai on the Israeli political spectrum.

When I received the call, I immediately returned it. Zakin told me his name was really Zakheim; he had changed it when he made *aliyah* in the 1930s. We marveled at the "might have beens" had the speaker of the Knesset and the American tormentor of Israel's military-industrial establishment sported identical names. We agreed to meet when I was next in Israel, but he died in mid-1987, and I never got to see my erstwhile namesake.

Prior to my departure from Israel, my friends Miki and Neal invited me to their home in Rehovot. Neal Hauser worked for IAI, and had written to me what it was like to work for that company while admitting to being a close friend of mine. "I inevitably get into quite an argument about you," he wrote. "Maybe 'argument' is the wrong word; 'heavy discussion' is better." Poor Neal; it wasn't easy for him at all. In fact, given the emotional level which the Lavi debate had attained and the nature of his job, it was downright courageous of him even to admit to our passing acquaintance, much less our friendship.

I hadn't seen Miki and Neal in a while, and I was happy to accept their invitation. Though I enjoyed my evening with them, it turned out that my trip to their home was more eventful than the dinner that followed. I hitched a ride with Tom Pickering, since he was scheduled to speak at a black-tie dinner at Rehovot's Weizmann Institute that evening. We were driving along on the Jerusalem road when our driver, Avraham, decided that he was not really interested in red lights that might slow his progress. So he pulled out a flashing blue light that he placed on the hood of our car and raced along the highway at speeds that would make even daredevil Israeli drivers quake in their boots. Pickering seemed to take it in stride, noting that Avraham often used the police light. Well, at an intersection not far from Jerusalem, Avraham, with lights flashing, ran a red light in front of an Israeli policeman. Israeli cops are not often held in high regard by their countrymen, and Avraham apparently subscribed to this view. But this particular policeman was not about to let Avraham get away, and he jumped on his motorbike and, siren screeching, chased us down the Jerusalem road. Avraham ignored the cop, and Pickering tried not to think about it. Avraham accelerated; the cop did the same. Stopping at red lights was now clearly out of the question, and Avraham, blue light flashing, and the cop, red light flashing and siren wailing, both pulled into the gate of the Weizmann Institute. The cop, clearly exhausted, and not a little bit annoyed, walked over to our window, which Avraham had rolled up. When he banged on the window, Avraham obliged by lowering it. *"Pratim,"* the man demanded—i.e., license and registration. Avraham's response was to shoot through the gate, which by now had been opened, and to pull up in front of the assembled crowd, all in black tie, and headed by the president of the institute. The policeman, undeterred, jumped on his motorbike and followed us to the receiving area.

"Pratim," he demanded again, as our driver turned off the ignition.

"Do you know who is in this car?" asked Avraham in Hebrew. My knowledge of Hebrew, which had improved considerably as a result of my frequent visits to Israel, was by now good enough to enable me to appreciate the true madness that often colors daily life in that country.

"I don't care who it is," was the reply, also in Hebrew.

"It is the American ambassador."

"I don't care if it is the president. *Pratim.*"

"Look, do you want to start an international incident?"

At this point Pickering intervened. "I am the American ambassador. My driver was following my instructions. I am responsible."

"Good," came the reply. "I will write him a ticket and you will tear it up."

All this time, the puzzled hosts stood at a respectful distance. After all, his excellency had not yet emerged from the car. Pickering now got out, and walked around the car to the policeman.

"I really think you are carrying this too far."

"I am only doing my job, just as you are doing yours."

Tom sighed heavily, said his good-bye to me, and moved toward his hosts. As I walked out of the institute, I turned back to see the policeman, the driver, and several men in black tie arguing heatedly. The cop was clearly not going to budge—and the alliance with America be damned. He got his way; after all, this was Israel.

15

We Finish the
Alternatives Study

Upon our return from Israel we found that, apart from reports in the trade press, the naval modernization study had attracted very little attention in the media and none at all on Capitol Hill. On the other hand, newsprint about the Lavi study continued to flow, much of it inaccurate.

By mid-October the Lavi alternatives had begun to take shape, but the contractors had not yet developed their cost and schedule estimates. Each of the three contractors was prepared to offer the Israelis several business and/or technical options relating to the four basic airframes that were being proposed. The General Dynamics "baseline" aircraft, from which its options derived, was the Peace Marble II version of the F-16 that it had already sold to Israel. GD assumed that Israel Aircraft Industries would be the prime contractor in Israel for licensed production of the plane, with airframe, radar, and engine production, as well as some subsystems production, all in Israel. GD offered eight variants of the basic F-16 configuration. They included night attack enhancements, incorporation of the Lavi's avionics systems, inclusion of other American avionics systems, propulsion enhancements, and training-related enhancements.

McDonnell Douglas offered four different alternatives—the AV-8B, the F/A-18, the F-15I, and the combination of F-15I and AV-8B. McDonnell put forward three options for the AV-8B that differed

not only in terms of the plane's configuration but also in the allocation of engineering integration, flight test, and aircraft delivery to
McDonnell and the Israeli firms. All of the five F-18 options, on the
other hand, addressed only configuration changes, while the F-15I
option incorporated a variety of improvements to the avionics, structure, and engine of the variant that had already been sold to the
Israelis.

Northrop also offered several options for its F-20 fighter. It gave
the Israelis choices between a sale through the foreign military sales
program or coproduction, between a single-seat and two-seat variant
of the plane, and between a plane with a combination of American
and Israeli avionics and one with purely Lavi avionics.

The contractors had done a remarkable job, considering the time
that had been available to them. Fortunately, thanks to the delay of
the Lavi's first flight, we had been able to slip the deadline for the
first draft of our study by eleven days to November 14. The contractors made it clear that they appreciated whatever schedule relief we
could give them.

By our next Lavi Alternatives Steering Group meeting, which
I convened on November 14, the day after arriving back from Israel,
we were able to give the contractors even more time, as the Lavi's
first flight slipped deeper into December. In any event, the air force
Systems Command had more information than it could usefully
insert into a briefing. Recognizing that I would not get to see all the
material that he had passed on to the air force, Vern Lee of General
Dynamics wrote to me directly on November 12 with what he considered to be important information about the F-16. He indicated
that GD had identified a delivery schedule for the F-16 that could
enable Israel to acquire the frontline fighters more quickly for less
money than the Lavi. This proved to be an important factor in the
Israelis' calculations when the fate of the Lavi was finally decided.

Late in November 1986 we received our first clear indication
that Israel's defense industries were in serious trouble, and that the
Lavi's cost and delivery schedule were in jeopardy. Writing in the
November 21 edition of *Ha'aretz*, Ze'ev Schiff summarized an MoD
report that had just been presented to Rabin,[1] which asserted that
Israel's major companies were losing money from the export deals
they had struck both because of the frozen dollar rate and because of
new wage agreements and bonuses. Schiff reported that IAI's hourly
rate had climbed to $34, roughly halfway between our own estimate

of $44 and IAI's $22 estimate. The company was anticipating losses of $100 million for 1986, the first time in many years that it was in the red. Such an increase in the wage rates would force IAI to slow down Lavi production in order to stay within Rabin's cap; in doing so, the alternative of acquiring F-16s more quickly became increasingly attractive.

It was also toward the end of November that I was able to circulate the draft report to my unofficial advisory team, Bob Levine of RAND, Herb Stein of AEI, and Stan Fischer of MIT. Charlie Wolf, dean of RAND's graduate school, had also agreed to provide us with comments, and my deputy Dick Smull briefed him as well. Other than Bob and Charlie, none of my advisers was any more prepared to be anything other than informal. The ongoing controversy over Lavi costs made them even less comfortable about a public link to our effort. It was easy for me to accede to their wishes. Their insights were excellent, and while I did not always follow their advice—which, as is often the case with economists, was not uniform anyway—it was always worth pondering.

Although it had been our original intention not to produce another report, the weight of the material forced us to draft one. It took us several weeks to organize the five hundred pages of data, and the first draft was completed on December 9. Once again we had to go through an arduous internal review process before we could submit anything to the Israelis.

The Israelis had a good idea of what we were about to release to them, since Ben-Joseph had been attending our steering group meetings. Eini also dropped by, on December 11, to learn firsthand of our progress. He asked to see the draft, but I turned down his request. He also reiterated that the Israelis preferred that we delay our visit until the new year. The reason for the latest delay was the same as the previous ones: the Lavi had yet to fly. We were determined to brief Rabin as soon as possible after the flight, however, and we had already cabled Embassy Tel Aviv a week before Eini's visit, requesting meetings with Rabin beginning on January 5, 1987. We also asked to meet with Peres, now foreign minister, and with the new finance minister, Moshe Nissim, who had replaced Yitzchak Modai, and whom I had not yet met.

Our work was virtually done, and the Pentagon was becoming restive about the Lavi's future. On December 15, a senior air force general, the director of air force international programs, told an industry audience that it was not in Israel's "best interests to continue the program."[2] It was the first time any U.S. official of authority had gone on the record to recommend that the Lavi be terminated. It was not to be the last time.

We received a number of comments to the draft during the middle of December, but none that fundamentally altered its thrust. The study was formally completed on December 21, though it was not to be released until after it had been briefed in Israel. As had been the case with the prior two studies, this one, entitled *The Lavi Aircraft: An Assessment of Alternative Programs*, had an unclassified executive summary that could be released to anyone.

The summary first laid out the ground rules that had been set for the contractors—the $475 million ceiling (in fiscal year 1984 dollars), the three-hundred-plane program, the objective of maximizing Israeli participation. It also specified that the alternatives assumed a start date of January 1, 1987, which everyone recognized was unrealistic, but which provided the most current comparison with the Lavi. The summary also noted that the quality of the cost estimates was "commensurate with Letters of Offers and Acceptance for Foreign Military Sales."[3]

The summary estimated that the Lavi program demanded an additional 96 million man-hours of work, if no new hires were assumed. This was to be the work-share target against which the various alternatives were to be assessed. The study pointed out that the contractors had put forward a total of nineteen options. Of these, we had chosen five for intensive analyses, with the remainder described in an appendix.

The summary then outlined each of the five alternatives in terms of flyaway cost, total program cost, funding delivery and profile, and total Israeli man-hours that each alternative absorbed. The first of the five to be reviewed was the McDonnell Douglas AV-8B. This alternative offered a low flyaway cost of $21.4 million per plane and a low total program cost of $7.4 billion, but the fewest Israeli man-hours, 39 million. The second alternative added fifty deep attack variants of the McDonnell Douglas F-15 to 250 AV-8Bs. The F-15's higher flyaway cost, $27.6 million, drove up the total program cost to $8.2 billion, but the Israeli work share also rose to 40 million man-hours.

The study put forward two approaches to an F-16 sale. The first called for licensed production in Israel of the Peace Marble version that had already been sold to the Israeli air force. The alternative had IAI serving as prime contractor for all but the center fuselage (as it had earlier indicated, GD had relinquished its work on the complex forward fuselage). The flyaway cost amounted to $14.6 million, lower than even Israel's estimate of the Lavi's flyaway costs, and the program cost totaled $4.7 billion. The study's summary pointed out that as many as thirty-six F-16s could be delivered annually, 50 percent more than the Lavi's maximum, and three times as many as the current Israeli budget allowed for. The study did not fully resolve the question of work share, because of uncertainty about the European Program Group. If a way could be devised to avoid their having to work on the F-16s, Israel would be able to realize as many as 55 million man-hours, or more than half projected for the Lavi, for a plane that would cost considerably less and be delivered three times as quickly.

The second F-16 alternative was more ambitious, but considerably more interesting. It provided not only for licensed production in Israel, but also for the insertion of Lavi avionics into the F-16 airframe. Since the F-16 was a larger plane, we did not anticipate major problems with what are termed "form, fit, and function," but there was certainly an integration challenge that had to be considered.

The flyaway cost of this alternative was somewhat higher than the Israeli estimate of the Lavi—$16.9 million compared to $15.2 million—but it had already become clear that the Israeli estimate was too low given the significant increase in IAI's wage rates. In any event, the $16 million figure was considerably lower than our own $22 million estimate for the Lavi.

The key to the second F-16 alternative was work share. Even if 15 percent of the work had to be allocated to the Europeans, a much less likely proposition given the radically different nature of the plane, Israel would still reap 68 million man-hours from the $5.8 billion program. Were the European partners not to get any work, 80 million man-hours, or 83 percent of the Lavi's level, would be available to Israel for a plane that was cheaper and that could be delivered at an annual rate of up to thirty units, more than twice the Lavi's expected delivery rate.

Our final alternative was not one from Northrop, as we had originally anticipated, but the McDonnell Douglas/Northrop F-18. This was the most expensive of the five, with a flyaway cost of $27 million and a program cost of about $9.5 billion. It also offered only

31 million man-hours of work for Israel. We included the option primarily to illustrate the variety of alternatives available to the Israelis. We elected not to include the F-20 because both Northrop and the Israelis seemed uncomfortable with the proposal.

Having set forth the alternatives, the summary—and the report—then noted that at the lower annual spending rate of $475 million, Israel could also move ahead with needed funding for follow-on support of the systems it already owned, with funding for the naval modernization program, and with the initiation of other needed programs such as assault and transport helicopters. Alternately, if Israel preferred to apply all of the Lavi's $550 million to aircraft alone, it could choose the F-16 or AV-8B alternatives, and, *in addition*, purchase twenty-four Peace Marble F-16s by 1991, by which time it could only have acquired eight Lavis. Such a move would bring further work to Israel, since the Peace Marble version of the plane also included a significant Israeli work share.

Finally, the summary noted that with the exception of the AV-8B/F-15 program, all of the alternatives would permit Israel to launch a major ATBM program within the $550 million cap, and several of them would permit as much as $140 million to be spent between 1987 and 1989. The reference to an ATBM program was the first of its kind in any official document available to the public. It laid the groundwork for American support for Israel's ATBM program, which, since the late 1980s, has centered on the Arrow missile, and has become increasingly critical to Israel's defense in the aftermath of Iraq's Scud attacks during the Persian Gulf War.

Having completed the report, we scheduled the obligatory set of briefings to pave the way for our forthcoming trip to Israel. We briefed Will Taft, the deputy secretary, on December 21, and Israel's Abraham Ben-Joseph the following day. Naturally, our most important briefing was to Caspar Weinberger; we met in his office on December 23. Dick Smull, my deputy, and Rich Higgins, SecDef's military assistant (who later was brutally murdered by Lebanese terrorists), were the only others in attendance. Weinberger praised the effort but wanted to know how a change from the Lavi would affect the strategic balance. I understood him to be asking whether Israel's security would suffer if the Lavi was dropped, but he hastened to correct me: "I mean, what will the Saudis think?" He had instinctively assumed that each of our alternatives was far superior in capability to the Lavi. Weinberger's concerns were not entirely misplaced; the air force had written to me two weeks earlier that the F-15I option involved capabilities far superior to those of the Lavi; American approval for the transfer of the F-15 had not yet been forthcoming. Ever solicitous of Riyadh's concerns in an administration that had done more for Israel than any of its predecessors, Weinberger did not want to spring another surprise on his Saudi friends. Bearing in mind the air force's admonition, I assured SecDef that the issue was primarily one of cost rather than capability, and that replacing the Lavi with any of the alternatives would not radically upgrade the capability of the Israeli air force. Weinberger seemed satisfied.

We also had one more major task to complete before the year's end. As is normally the case with any GAO report, we had been handed the draft GAO Lavi report on November 26 for comment prior to its release. As Derek Vander Schaaf, the deputy inspector general, had accurately forecast, the report yielded estimates somewhat closer to the Israelis' than to ours, but argued that the program was unaffordable given projected constraints on both our budget and that of Israel. While we were pleased with that conclusion, however, we still had a number of methodological and factual differences with the GAO. We were trying to complete our own Lavi report at the same time, and the burden of completing both tasks fell on the Air Staff and on Captain Linda Hardy in particular.

This was not the best time of year to produce much work of any kind. The federal government's holiday season, like that of retail merchants, seems to begin earlier with each passing year, and by the third week of December the Pentagon's rings and corridors echoed with the noise of an endless round of office parties. Still, the air force recognized the importance of producing a strong rebuttal to the GAO, and Linda pressed on, giving no evidence of being distracted by the season's festivities.

We completed our review of the GAO report at virtually the same time as we received the Lavi study back from the Pentagon printer. We immediately met with the GAO staff (on the same day as we briefed SecDef) to convey our evaluation of their study. Essentially we differed over the details, but agreed to the GAO's principal findings regarding Israeli funding and cash flow requirements. We also felt that the recent delays in the prototype flight schedule, the collapse of the Bet Shemesh engine factory, and the Israeli reports of wage rate increases vindicated our higher estimates.

Although the GAO's analysis had yet to see the light of day, the

Israelis seemed cocky about its findings. Eini had asked me when the report would appear and seemed positively buoyant when he referred to it. By December 12 the State Department, which had also received the report for comment, had responded to the GAO that it had none. State's prompt response made it seem that DoD was stalling. A leak to *Defense News* in late December accused me of deliberately trying to prevent the report's release because the GAO had challenged our cost estimates.[4] Like so many other of the allegations that emanated in Israel, however, this one was completely untrue.

The Israelis were in fact playing a double game. On the one hand, they insisted that they wanted the GAO report released before ours, so that it could detract from whatever we had to say. "An Israeli source who asked not to be named" was quoted in *Defense Week*'s January 5 issue as observing, "It seems strange that Zakheim is in Israel this week and the report has not been released. . . . it's clear that he's held it up."[5] The Israelis apparently assumed, quite incorrectly, that our cost estimates would be discredited if the GAO's estimates were not identical to their own. Either Eini and his supporters were not aware of the GAO's true "bottom line" or they simply undervalued its import.

Nevertheless, even as they tried to score publicity points from the delay, it was the Israelis themselves who were holding up the report. As had been the case with respect to our own Lavi studies, the Israelis insisted that the GAO effort be classified. Since the GAO also publishes unclassified versions of its reports, it was necessary that the DoD inspector general review it in accordance with security classification procedures. The IG staff was as caught up in the holiday spirit as everyone else in the Pentagon; it was therefore not until January 2 that we were able to transmit fourteen pages of detailed comments, as well as the results of our security review, to the GAO.

Even as we recognized that the appearance of the GAO report might dilute the force of our estimates, the economic realities were beginning to overshadow paper estimates from any source. Late in December the Jerusalem Television Service revealed yet more growth in IAI's wage rates. They now stood at $37 an hour, several dollars higher than Schiff had reported the previous month. Even more important, IAI had been forced to reduce labor on the Lavi by 25 percent, resulting in a delay in production to 1991. Even then, the first planes to enter the Israeli air force would not be fully equipped with all the subsystems originally intended for the aircraft. Finally, the report noted, unless Rabin's $550 million ceiling was lifted, IAI could deliver no more than ten planes annually, resulting in a program that would span two full decades or more.[6]

IAI's failure to keep its wage rates down was the beginning of the end for the Lavi program. No amount of polemic could prevent the program delays, or the cost increases that would accompany them. Israel's defense budget and our security assistance simply could not pay IAI's bills. Our alternatives report took on far greater urgency as a result.

I had one last piece of unfinished business with respect to the naval modernization study that I hoped to complete before I left for Israel. I had promised the Israeli navy that I would try to get SecDef's support to cover the shortfall between our recommended offset for 75 percent of the submarine work done in Europe and the total cost of that work. I hoped that Weinberger would agree that the remaining 25 percent—less than $50 million—be covered by offshore procurement funds.

Fred Iklé supported my proposal, which I sent to Weinberger as an action memo for his approval. It was returned to me on the morning of New Year's Eve with the note "Let's talk." Cap was already out of town for the weekend, but he phoned from Maine. I put the case to him again, but he was noncommittal. He first wanted to be sure that the Israelis would terminate the Lavi before he made any decision on funding the submarines. Still, he did not reject the idea outright, and I took that as a positive sign. The way was now clear for our trip to Israel.

IAI kept its promise to its supporters: the Lavi flew in 1986. It took off at 1:31 on December 31, only ten and a half hours before the new year. The flight lasted twenty-six minutes. It was nine months late.

Few members of the media were invited to witness the flight or attend the ceremonies that followed. All material supplied to the press was carefully controlled. It was, as one reporter put it, "a preemptive strike" that was "intended to create . . . a suitable psychological effect."[7]

Moshe Arens was on hand for the takeoff, as was Air Force Commander Amos Lapidot. But Chief of Staff Levy did not show up at all, while Yitzchak Rabin missed the flight and only caught the reception that followed. Speaking to the program's supporters at IAI, Rabin acknowledged their technological achievement but added that the program still faced many obstacles and noted that funding was being provided solely by the United States. He again stressed the obstacles in the way of the program when he spoke to the Knesset later that afternoon and for the first time commented that the Lavi's future depended on both Israel and the United States. It was a performance reminiscent of a baseball owner's vote of confidence in the manager the day before the manager is fired.

Menachem Eini put a different spin on the flight, of course. Sounding like a gunslinger challenging his foe to a shootout at high noon, he announced that "we are waiting for Zakheim."[8] Others noted that it was one thing to criticize the Lavi when it was a "paper airplane," but quite another to take on a real system that had actually performed operationally.

Though everyone celebrated the demonstration of Israel's technical prowess that the Lavi represented, the flight did little to get partisans in the debate to change their minds. Many commentators stressed the high cost of the plane, not only in light of the recent revelations of labor rates at IAI, but because it had become clear that the Israeli air force would purchase fewer than three hundred aircraft.[9] It was now estimated that the unit cost of the Lavi would approach $50 million, and that, because of the program's schedule delays, additional American aircraft would still have to be purchased as "gap fillers" to replace Israel's older F-15 and F-16 models.[10]

All of the major alternatives had leaked to the media by the time we arrived in Israel; licensed production of the F-16 in Israel emerged in the press as the most interesting competitor to the Lavi.[11] Although Eini claimed at every opportunity that our alternatives were neither more interesting nor less expensive than the Lavi, it was becoming increasingly clear that fewer and fewer leaders of the defense establishment were taking him seriously. Several reports noted that the defense ministry appeared to be gearing up to kill the program. The Lavi episode was soon to reach its climax.

We arrived in Israel four days after the Lavi had finally flown. I was joined by Dick Smull, Len Zuza of OMB, Jim Stark of the NSC staff, and several U.S. Air Force officers. Although my previous visits to Israel had been marked by press attention that seemed overwhelming, nothing prepared me for the reception I encountered on this latest Lavi-related shuttle. Eini was obviously not the only one who was waiting for me. I was mobbed by reporters from the moment I came down the steps of our air force jet; they wanted to hear me say anything, however meaningless it might be. Exasperated, and tired from the journey, I remarked that the weather seemed fine. This earthshaking observation became a headline in the next day's edition of the mass-circulation *Yediot Achronot*. When the country was drenched in a downpour that day, it only proved to the Israeli public that I was not to be trusted.

As I walked into my hotel room at the Tel Aviv Hilton—the hotel staff had been instructed not to reveal my room number—the phone rang. It was a reporter; he had watched me walk into the hotel and go up the elevator. When I checked in with our embassy, I was given nearly a dozen messages, from acquaintances dating back to high school and through college and grad school, and, of course, from my "relatives." My every move was reported on the hourly broadcasts of Kol Israel radio, which I heard when I arrived at the hotel, and later the next morning, as we made the short drive from the Hilton to the U.S. embassy and then to the defense ministry. I began to feel that this was not another official visit; it was a three-ring circus.

My arrival was greeted by a new onslaught from the plane's proponents. I had asked to meet with Tom Pickering before we saw any of the Israelis, and when I arrived at our embassy in Tel Aviv on Monday, January 5, the attaché's office first handed me a thick summary of media reports of the previous two days to review. It was an eye-opener. A source at IAI had told *Ma'ariv* that I was anti-Zionist, and alleged that from what I said it would seem that I "preferred to see Israeli engineers driving taxis in New York rather than making a first-class aircraft in Israel." By comparison, Eini's remarks to Israeli radio seemed rather mild. The program manager asserted that the GAO had already proved my original estimates incorrect, while he was happy to see that his estimates "were good and accurate." He stated that we were bringing no novel alternatives; the only one that was "relevant" was the F-16, but "a marriage between that aircraft and the Lavi avionics won't do."

Meanwhile, over the weekend, Prime Minister Yitzchak Shamir had insisted to a visiting delegation of British Jews that the project was a "matter of life and death" for Israeli technology. IAI's new chairman predicted that it would take a "generation" to undo the damage caused by cancellation. A group of retired pilots, senior air force officers, and former IAI officials announced the formation of a "lobby" of sorts to press for the Lavi. If that were not enough, *Davar* reported that the embassy was going to arrange a meeting between me and "local anti-Lavi reporters." To save the day, the report asserted, "Israel's regular aviation correspondents . . . asked the embassy to let them join the meeting."[12]

Virtually every newspaper also announced my impending meeting with Rabin in front-page banner headlines. *Ma'ariv* headlined a quote from a "very senior defense official" who said there was nothing to the Lavi alternatives and pointed out that a government that could not even bring itself to close the Bet Shemesh factory because seven hundred jobs were at stake simply could not be expected to terminate a program affecting ten thousand workers.[13] *Davar* was full of articles impugning our purported selective leaks to anti-Lavi journalists of both the alternatives study and the GAO report. Presumably this was the same bunch of media types that I was planning to brief privately. Other articles announced that Abraham Ben-Joseph had reported back from the New York purchasing office that there was nothing to our proposals, while former MoD director general Mendy Meron insisted on television that it would take a few months to evaluate the alternatives but that in any event the program should proceed unchanged. All reports made it clear that the government would take no action while I was in Israel; that Rabin would still want to discuss the matter with Weinberger, but would refer the issue to the cabinet yet again on the week after my departure.

After glancing through the press reports, I previewed our briefing in Tom Pickering's office. Tom liked what he saw, but warned me that Menachem Eini and his friends would do everything they could to discredit what we had to say. He urged me to stay cool in the face of what was sure to be intense provocation. Tom's advice had always been sound; it did not fail me this time either.

I was pleased that Pickering was once again going to accompany me to Rabin. I did not anticipate an easy morning. We proceeded from the embassy, which is located in the heart of the Tel Aviv beachfront hotel strip, to the Hakirya complex in north Tel Aviv. We began with a short meeting with Rabin; Eini and Shimon Hefetz, Rabin's aide, were also present. Tom Pickering brought Rabin another note from Caspar Weinberger. Tom told the defense minister that an identical note, signed by George Shultz, was addressed to Prime Minister Yitzhak Shamir. The two American secretaries asked that the Israeli government give our alternatives most serious consideration and made it clear, at least by implication, that they were unhappy with the Lavi. Because the note had been signed by both men, there could be no mistaking that Shultz was taking a very hard line regarding the Lavi's future prospects. His reputation both as a friend of Israel and as an economist who had done so much to help Israel's economic recovery made his advocacy of the alternatives even more compelling.

Following the introductory discussion, we assembled in the MoD conference room that adjoined Rabin's office. David Ivri, Eini, and Chief of Staff Levy were there, as well as innumerable colonels and other officials. Tom Pickering also sat in. The room was packed.

In line with the old Pentagon adage to "tell 'em what you're going to tell 'em, then tell 'em, then tell 'em what you've just told 'em," we began with an overview of what was to follow. We then described what we termed as "the problem": rising costs, delays in the flight and the closing of Bet Shemesh, the impact on Israel's military program, and Rabin's own $550 million cap. We then summarized our terms of reference, emphasizing that we were addressing a three-hundred-plane program—the Lavi's original force goal—rather than the currently proposed truncated Lavi production run. We also noted not only our concern to maximize Israeli production, but the role of Israeli observers in the activities of our interagency steering group.

Rabin did not ask many questions as we turned to our evaluation of the Lavi's current status. We noted that our estimate allowed for 96 million man-hours remaining on the plane, but that the production plan was incomplete, while the tooling was "soft"—that is, especially made for the prototype and not geared to serial production. This was likely to add more work to the schedule.

We completed this portion of the briefing by emphasizing that the Lavi program assumed an optimized requirement for spares, which was likely to increase (and thereby increase costs), and that there was limited integrated logistics support. Eini was inclined to interrupt me at every turn, and squirmed in his seat as I repeated that the plane was "unproven," but Rabin kept Eini firmly under control.

I then turned to the alternatives, describing them as we had in the unclassified summary of the report. Having proceeded through each of the five, I outlined the additional programs, such as the ATBM program or the purchase of twenty-four Peace Marble F-16s, that adoption of one of the alternatives made possible. Rabin was not at all impressed by the ATBM suggestion. He dismissed the entire idea with a grunt, indicating plainly that he saw no need for an Israeli anti–tactical ballistic missile program. Within two years he had changed his tune. Significantly.

I concluded that all of our alternatives generally met the mission requirements as the Israelis had specified them. Reiterating that the Lavi's costs were at best the subject of debate, I noted that the alternatives not only would provide work for Israelis, but would yield resources for other projects. I then reemphasized that the alternatives were all operational aircraft, as opposed to the unproven Lavi, and that they created less pressure on our security assistance budget, which, I assured Rabin, would not exceed $1.8 billion.

The briefing had run on for about ninety minutes when Rabin brought it to a close. He agreed that we should meet again the day after next to review any outstanding matters. Once we moved out of Rabin's office suite, it was left to Eini and Ivri, but especially Eini, to pepper us with questions about the report. He was decidedly unfriendly. He also stated that he needed time to digest its contents.

The meetings at Hakirya took up much of the day. We left several copies of our briefing with Ivri and his staff and proceeded to wade past the sea of reporters who mobbed the entrance to the building. I arrived back at my room utterly exhausted. Each meeting had involved not only detailed discussion of our alternatives but sipping innumerable cups of Israeli (i.e., Turkish) coffee and munching on stacks of biscuits. By day's end I had consumed far too much of both, and, combined with the adrenaline that coursed through my system whenever I came to Israel, I was running on super high octane. If I got three hours of sleep that night, it was a lot.

The next morning we stopped off at the embassy to take in reports of the previous day's events before traveling to Jerusalem to brief Shimon Peres. All newspapers had given the meetings front-page coverage, and most editorialized on the Lavi as well. Some papers went even further. *Hadashot*, for example, had a two-page inside spread that detailed each of the alternatives, complete with photos of the four aircraft involved.

Unnamed "senior sources" at the defense ministry were cited everywhere. Some expressed "strong irritation" at our supposed attempts to interfere with IDF policies. Others supposedly were backing away from the Lavi to support the purchase of F-16 C/D models. Israeli journalists are a mixed bunch; some are both dedicated and thorough, others subordinate thoroughness to their own biases. Many are known to rely on a single source. Were the pro-Lavi sources more senior, less senior, or as senior as the antis? How many were there in each camp? It was hard to tell.

What was clear was that the defense establishment was now openly and bitterly divided on the fate of the project. Moreover, that division extended to the alternatives. We had, to a great degree, already succeeded in our mission, regardless of whether any one of our alternatives was chosen. The issue was no longer a matter for IAI, or even the defense ministry, to decide. It was no longer an obscure issue chiefly of interest to economic analysts. As one writer put it, the issue had passed from the analysts to the politicians, who had to weigh conflicting pressures not only among the services—which was a matter for Rabin—but from both civilian and military industries, depending on whether or not they benefited from the project. And then there were the other ministries, led by the finance ministry, which, feeling that the Lavi forces were on the defensive, positively salivated at the prospect of either grabbing those funds for their own programs or, in the case of the finance ministry, simply for better control of the budget.

It was also becoming widely recognized that, however unrealistic it might be, there was really but one way that Israel could avoid having to choose between the Lavi and its alternatives. Israel should appeal to the U.S. president and to Congress over the head of the Pentagon, as it had in the past. As the left-leaning daily *Al Hamishmar* put it, "The fact that the U.S. Administration and Congress have been allocating money to the Lavi project makes them as accountable as Israel for that investment."[14]

In a sense the newspaper was correct. The Congress had never taken a tough stand on the Lavi. The Israelis could argue with some justification that it was too late for the Pentagon to complain. It should have filed those complaints, and undertaken all its studies, much earlier, directing them to the Congress, not to the MoD.

No doubt this argument was laced with cynicism. Virtually all informed Israelis recognized that congressional pliability on this or any policy relating to Israel did not necessarily connote sincere agreement with that policy. It was as much a reflection of domestic U.S. politics, most notably of deference to AIPAC and to Israel's other Washington allies on all matters relating to Israel. Nevertheless, Israel could not be blamed for its success on the Hill. Whatever the reasons, the fact remained that legislation relating to the Lavi, without which the program could never have survived, was a signal of official acquiescence on the part of the U.S. government. It was a tough argument to counter, and we didn't even begin to try to do so. Instead, we attempted to keep the debate firmly planted in the mine-laden soil of Israeli domestic politics, rather than permit it to return once again to the American domestic political scene.

We also found that the debate had ranged beyond both Israel and Washington to a new arena: the Middle East conflict. The Arabic-language paper *Ad-Dustour* attacked us as well, characterizing our effort to get Israel to drop the Lavi as "a present from the U.S. to Israel and a form of declared war against the Arabs." And, to be sure, we *were* telling the Israelis that by choosing an alternative to the Lavi they would better preserve their qualitative military edge over their enemies.

It is a well-known Washington adage that if a policy is attacked from two opposite directions, it probably is good policy. Not that we needed any more convincing, but the fact that the Arab press began to oppose our effort, even as Moshe Arens and the right-wing Likud did as well, reinforced our belief that we probably were right on the money.

After taking in the morning press, and the first of the day's cups of coffee, we helicoptered to Jerusalem, landing at the Knesset helipad and making the five-minute drive to the foreign ministry. Foreign Minister Shimon Peres was still strongly advocating Israel's commitment to high technology as embodied by the Lavi. But he had few incentives to go to the mat on the issue. With the center of Labor no longer fully in support of the program, despite the agitation of the IAI workers, Peres, ever sensitive to the ebbs and flows of opinion within his party, was not about to let himself be seen as the Lavi's champion. In any event, Arens had long since arrogated that role to himself. The minister without portfolio was continuing to do his utmost both to cast the project in nationalist terms and to ensure that there were no strays from the Likud herd.

Once again, I was fortunate to have Tom Pickering with me for our Jerusalem briefings. When we arrived at the foreign ministry, we first met Nimrod Novick, who, despite his relative youth, was Peres's veteran political and media adviser. Nimrod's presence meant that the press was not too far behind. Sure enough, and in contrast to the scene at Hakirya, which was abuzz with press outside the main building but quiet within, Nimrod let the press into the foreign ministry's main conference room for five minutes of picture-taking. They then were ushered out of the room.

Peres made his obligatory opening statement in support of the Lavi, qualifying it by indicating a willingness to entertain the alternatives if they would preserve Israel's technological base. I responded by acknowledging the technical feat that the flight of the Lavi represented, but pointed out that the test flight in no way affected our conclusions or the urgency of the alternatives. We then went through the briefing again, as we would a half-dozen times before our departure. Peres thanked us and said he needed to study the details. After we waded our way through the press horde outside the ministry, Peres informed the reporters that he still supported the project. But he then went on to repeat what he had just told us, that he needed to study the details, and also added that the United States was not applying pressure on Israel, but was only offering suggestions.

Even as we were briefing Peres, the repercussions of our meeting with Rabin, and perhaps even more important of the letter he had received from Weinberger, were being felt nearby in the Knesset. Although he had cautioned his staff just six months earlier not to criticize the project, IDF Chief of Staff Moshe Levy did just that before the Knesset's foreign affairs and defense committee. Reverting to his characteristic bluntness, Levy told the legislators that had the military been operating from a clean slate in 1987, there was no way that the program would have been initiated. He added that it made no sense to support a wounded project, and that it would be impossible to support the project properly unless funding was found outside the IDF's budget. The only way the Lavi could be conceived of as a "national" project, in line with the claims of its supporters, was if its funding was significantly increased.

Levy's statement marked a breakthrough of sorts. Though Ivri would remain a diehard supporter of the Lavi through the bitter end, and even years later, Levy's remarks both reflected the increasing uneasiness of the military with the project and effectively opened the door to new anti-Lavi leaks from the IDF to a press that was all too eager to print them. The opposition to the plane was gathering momentum even as we continued on our round of briefings; in fact, it appeared that the majority of Israel's top military leaders were now opposed to the aircraft.

Before leaving the foreign ministry I met briefly with Abrasha Tamir, the dovish former general who was Peres's director general at the foreign ministry. Unlike Peres himself, but very much like Amnon Neubach, Tamir had absolutely nothing good to say about the Lavi. He encouraged me to carry on, despite the obstacles; he was convinced the plane was a waste of money.

We had a few hours' break before our next stop at the finance ministry, and I headed in the direction of Strauss Street, which is near the very Orthodox neighborhood of Meah Shearim, and not far from the government offices. The embassy drivers who took me around to my various appointments were by now fully aware my private interests were somewhat different from those of most of the people they chauffeured. When staying in Jerusalem I would often return to my hotel via nearby Meah Shearim, where the best religious bookstores could be found.

On this occasion, I stopped at the Hasofer scribe's shop, where one can purchase a wide variety of ritual articles. Among these are tefillin (phylacteries), small hollow boxes containing four extracts from the Bible, written on parchment, that are worn on the arm and forehead during daily prayers, and mezuzahs, small receptacles containing two Biblical extracts, likewise on parchment, that are placed on the upper right doorpost of a home's entrances and important rooms therein. I bought a pair of tefillin for my son Chaim, who was to be bar mitzvah later in the year (one only wears them from the age of thirteen onward). My driver decided that he too needed something; he had not replaced some old mezuzahs, and did so as I made my own purchase.

We made one more stop on the way back; I wanted to visit the world-famous Mirer Yeshiva in Jerusalem. My relationship with the yeshiva was a bit convoluted. When my father escaped from Lithuania just after the Soviets occupied the country, and only three months before Hitler's surprise attack on Stalin's forces, he had signed the exit visa applications for the members of the yeshiva and had traveled with the group across the USSR, looking after their legal interests in Shanghai before emigrating to the United States after the war's end. He had settled in a part of Brooklyn that was but three blocks from the branch of the yeshiva that had been established in the borough, and regularly attended its Sabbath services, which I attended with him until I was old enough to frequent another synagogue.

I had always been curious about the Jerusalem branch, and so took the opportunity to inquire as to whether anyone there remembered my dad. I was delighted to discover that was indeed the case, and pleasantly surprised at the warm reception that I received. It was made clear to me that the *Eidah Charedis*—the religious community—had no truck with the Lavi, which was seen as a drain on resources that otherwise might have been allocated to needs such as religious education. My opposition to the Lavi therefore very much dovetailed with their views, and the rabbis were gracious to the point of my being escorted onto the street by one of the institution's two deans—termed the *rosh yeshiva*—when I took my leave.

My driver, of course, took this all in. Although he was not a particularly religious man, he told me as we drove back the mezuzahs had weighed heavily on his mind for some time and he was pleased that he had finally bought them. I guess my religious interests had gotten to him.

Our final meeting that day was at the finance ministry, which was now being led by Moshe Nissim, who, like his predecessor Yitzchak Modai, was of the faction of Likud that had once been the Liberal Party. Nissim first asked to see me alone. I came in to find a man as different from Modai as he could possibly be. For a start, he was more rotund than the tall, slender Modai. He also wore a huge black yarmulke on his head. I had not realized that he was an Orthodox Jew.

Nissim just stared at me for a while, and told me in Hebrew that he wanted to get an impression of me. We chatted inconsequentially, even as something in the back of my mind kept telling me that there was more that we could have discussed—and it wasn't the Lavi.

We briefed Nissim and his staff for about seventy-five minutes. The finance ministry had always been friendly territory, and it was good to see its director general, Emmanuel Sharon, again. Nissim did not openly state his opposition to the Lavi, but kept emphasizing the need for cost constraints given the pressure on Israel's budgets. I left his office uncertain whether I had gained an ally to replace Modai. The latter, perhaps in response to pressure from Moshe Arens, perhaps because he no longer was under the influence of the finance ministry staff, had reversed himself and had adopted a pro-Lavi stance.

As I was leaving Nissim's outer office I remarked to one of his aides that he seemed to wear a large yarmulke. My comment prompted some surprise. Didn't I realize that his father was HaRav

Nissim? I could have kicked myself. Rabbi Nissim had been the Sephardi chief rabbi of Israel, a man revered for both his piety and his scholarship. I would have loved to have returned to the minister's office to chat about his father, about Jewish law, about mutual acquaintances in the world of Jewish learning that had little to do with fighter jets. Instead I had to move on, as did he, and I was reduced to writing him a brief note, in the hope that we could take up our acquaintance at some future time.

The next morning I began my day as we had the day before, by scanning the newspapers. There had been no letdown in the breadth and scope of the coverage; if anything, it seemed to have expanded. One interesting development was that the ministry of defense evidently had recalculated the loss of jobs to Israeli industry if the Lavi was dropped in favor of the F-16. *Ha'aretz* reported MoD's new estimate that five thousand jobs would be lost, instead of the ten thousand trumpeted by IAI.[15] There were also other reports that IAI had exaggerated the number of people currently working on the Lavi project.

We drove to Jerusalem to brief Prime Minister Shamir in his offices across the road from the Knesset. We arrived somewhat early, which afforded me the opportunity of an impromptu meeting with Ezer Weitzmann, who, like Moshe Arens, was a minister without portfolio. Weitzmann's office was in the building that also housed the prime minister's office. I had never met Weitzmann, the former defense minister who could lay as much claim to being the father of the Lavi as did Moshe Arens, but he warmly welcomed me into his office. We chatted about the project as I quickly walked him through some of our slides.

Weitzmann would delight in shocking people who met him for the first time. Though subsequently, as president of Israel, he forced himself (no one else forces Ezer to do anything) to be more of a diplomat, he certainly was nothing of the sort when we met. He referred to the Lavi and its supporters in words that rarely were longer than four letters, unless used in the participial form. The fellow is brilliant and has been blessed with an Irishman's gift of gab and a double dollop of Irish charm. I lapped it up.

We moved down the hall to the prime minister's suite of offices. Although this was an official visit, not a personal one, Shamir was friendly as usual. I showed only a few of our charts; cost analysis is not his bag, and he couldn't avoid dozing off. When we finished, he said that this was a matter on which he would have to consult his "good friend Arens, who specializes in these matters." He made it clear, though, that he was not about to back away from his recent public statements about the importance of the program to Israel's technological health. He also emphasized that the decision would be Israel's alone to make, not ours. That was about what I had expected, and I was pleased that our substantive disagreements had in no way marred our personal interface. That was more than could be said about my relationship with the "good friend" Moshe Arens.

Our next stop was the Knesset, an imposing rectangular structure whose inner walls are adorned by Chagall paintings. Our destination was the finance committee, which was chaired by Rabbi Avraham Schapira, a wealthy Hasid and a leading light in the very Orthodox Aguda Party. My team did not realize who Schapira was, and I was amused by the reactions of many on our side, particularly the air force staff officers, who perhaps were used to congressional hearings, when the corpulent, red-faced Schapira, dressed in black caftan, with long earlocks *(peyot)* and a wide-brimmed black hat, marched into the room and took his seat at the head of the committee table.

Schapira was no slouch when it came to finances, as befit a man who had made his fortune in the export-import trade. He recognized the force of our case and was very sympathetic to our alternatives. No doubt he was also influenced by his party and religious role; as I had found the day before, the religious parties, always hungering for more funds for their educational and social institutions, rarely objected to reductions in the defense budget if social programs stood to benefit therefrom.

After our briefing we stopped at the Knesset cafeteria, meeting a few more "MKs" before seeing Rabin in his Knesset office for a wrap-up session. He thanked us for our contribution and said it would take some time before his specialists could digest all of the details that were in our report. We discussed the next step, which was to have the American contractors come to Israel to brief their various alternatives to the MoD and to IAI. Eini wanted to hold discussions with the General Dynamics team, and it was agreed that representatives of General Dynamics, Westinghouse (which produced the F-16 radar), and General Electric (which produced the engine) would arrive on January 13; I asked Dick Smull to head the group. The Israelis did not indicate when they wished to hear from McDonnell Douglas.

As we walked out Rabin's door, Tom Pickering and I discussed the possibility that I brief Arens, since we would be briefing virtually every other key player in the Lavi matter. I asked Rabin's military aide, Shimon Hefetz, what he thought. He agreed that it was a good idea, and told me to wait while he walked back into Rabin's office. He came out saying that the defense minister also agreed; Shimon suggested I phone Arens then and there.

Arens was in his office to take the call, and it was in his office that he said he would be happy to receive the briefing. The problem was that his office was in East Jerusalem, and it was U.S. government policy not to conduct official business with Israel in East Jerusalem. Arens knew this, of course, but I explained it to him anyway, offering to go anywhere in Israel to meet him (I used the Biblical phrase "from Dan to Beersheba"). I covered the mouthpiece to consult with Tom Pickering, who felt strongly that we could not offer a briefing in East Jerusalem. Arens, however, was not interested in being briefed anywhere but in East Jerusalem, and, after I remonstrated with him that I was in an impossible position, he politely said, in effect, thanks but no thanks. Tom said that at least we had made the offer.[16]

It was finally time to head back to Tel Aviv. As I was walking out the door, a Hasid came up to me and blessed me, adding the Talmudic dictum *Al tehi birchas hedyot kal beeynecha*, "Do not belittle the blessing of an ordinary person." I was delighted to accept his blessing; it made a welcome change from some of the invective that had flown my way.

Having completed what seemed by now to be an interminable round of briefings and meetings, we returned to the embassy to face the press. I expected the press conference to last about an hour. I suppose it did not extend much past the hour, but it too felt endless. I tried my best to convey that the decision was Israel's, not ours; that there would be no more security assistance to help Israel out of its financial bind; that we would not buy the Lavi; that we would not be helpful if Israel tried to export it to other countries. Although I was pressed on the matter of the GAO report, I felt obligated not to divulge or even hint at its contents. I did make clear that the report, like many parts of our own cost analysis of the Lavi, had been classified at the request of the Israelis, and not because we wished to hide anything.

By now I was exhausted, but there was one more cocktail party to attend. I had to go; it was hosted by Menachem Eini. I was asked to say a few words and proceeded to thank my Israeli counterparts in both Hebrew and English, and voiced the hope that we could cooperate to see this effort through its conclusion. It was a vain hope; but it sounded good.

I was pleased to find the next day that the press corps had generally been quite accurate in reporting what I had said. More important, the anti-Lavi drumbeat was continuing, both reflecting and nourishing the widening schisms within the military establishment over the project. *Davar* reported that Rabin's deputy minister had refuted IAI's claim of massive unemployment if the Lavi was terminated and had asserted that some of our alternatives would actually lead to *new hires* for engineers.[17] The *Yediot Achronot* correspondent, who had attended the press conference and asked several penetrating questions, reported that I had not even addressed the real alternatives to the Lavi—and the reason it was in trouble with the military. Were the project to continue, it would come at the expense of land forces modernization (in a military still dominated by the land forces), as well as of the acquisition of critical stand-off and electronic support systems. "The Lavi," he wrote, "came too late and too expensively" to meet the new demands upon the military.[18] And Reuven Pedatzur of *Ha'aretz* reported the anger of senior air force officials at the decision to have the Lavi project office review our alternatives. These officers were claiming that there was a clear conflict of interest on the part of the reviewers.[19]

Pedatzur also noted senior air force leaders' concern that the possibility of acquiring F-16s without industrial embellishments not be dismissed in the welter of alternatives that I had put forward. It was this kind of report that was certain to keep the Lavi in the headlines well after both our departure and that of Dick Smull's contractor team the following week.

16

Exiting DoD

The trip to Israel marked the culmination of my effort on the Lavi. I expected that the rest of my duties with respect to the project would essentially be a mopping-up operation. I anticipated that there would be more interviews and backgrounders to the press, more briefings on the Hill, more speeches, more discussions every time a senior Israeli official came to Washington. Nevertheless, I felt that the critical part of my task—developing the analysis, presenting it to Israel's decision-makers, and ensuring that the Lavi became a matter of debate within that country's policy elite and body politic—had successfully been completed. The *Jerusalem Post* noted shortly after our departure that "by the time Zakheim and his team left Israel . . . the Lavi's future had never looked bleaker."[1] The next real step had to be taken by the government of Israel.

Beginning in early November 1986, I had decided for a variety of personal reasons that I wanted to leave the government. Government service had been exhilarating, but, like many who have served in government, I was finding it increasingly more difficult to pay the bills and provide for my family's future. I was sending my sons to private Jewish school, and they soon would enter high school, where the fees would be even higher. College was just around the corner for my eldest son, Chaim, and I could anticipate having to pay college tuition bills for two of my boys at the same time in the not-too-distant future.

I also had to face the reality that I was a political appointee, and had already served far longer in that capacity than the usual political type. That had much to do with Caspar Weinberger, who was still to serve another year, and who set the tone for the department. Many of our top officials served six or more years: Fred Iklé and Rich Armitage, to name but two in the policy group. But I felt it was time to move on.

Much to my surprise, however, between the January trip and my departure at the end of March the Lavi continued to dominate my time. We had succeeded in making the Lavi the single most controversial defense issue in Israel and among the most controversial issues overall in that controversy-racked country. As a result, because I was the most visible American exponent of the plane's extinction, I remained in the center of the political hurricane that I had generated.

The momentum for change had continued to accelerate as the American press reported the growing doubts of the Israeli military. Air Force Commander Lapidot's confession the day after my departure that "I love the Lavi but can live without it" and IDF Chief of Staff Moshe Levy's testimony before the Knesset committee drew special attention. Unlike their respective deputies, Bin Nun in the air force and Dan Shomron on the general staff, Lapidot and Levy had been vocal proponents of the program only a year before.[2] Equally telling were reports of a split in the Lavi camp that emerged in late January. At a meeting in Rabin's office to discuss the findings of the committee that was reviewing our alternatives, Eini's Lavi project directorate differed with IAI over the cost and schedule of the airplane, with the project office estimating that the Lavi could cost as much as $18 million, which we knew was roughly the GAO's estimate. A follow-up press report asserted that the Lavi program office had calculated that the project could only make economic sense if three hundred aircraft were purchased, but that the Israeli air force could not absorb more than one hundred planes. Some military officials accused the program office of deliberately basing all calculations on a three-hundred-unit buy (as we had for our alternatives) in order to obscure the true cost impact of the program.[3]

Late in January it also became clear that, much as we had predicted, the Lavi was becoming heavier. It was announced that the Lavi required a wing redesign because Israeli engineers had added four thousand pounds to the plane's gross weight. While the Lavi program office insisted that the change would have little impact on cost, it was obvious that the weight increase of more than 10 percent would drive up costs, while the redesign, with its impact on the plane's stability, was certain to delay the production schedule.

The growing Israeli din over the project had a noticeable effect on the American Jewish leadership. The more it absorbed the criticism that Levy and others in the military were leveling against the plane, the more noticeably silent it became. This in turn made it easier for the Congress to adopt a more neutral stance, while the administration became increasingly vociferous in its opposition to the plane. And the pressure from people like George Shultz made it easier for politicians who quietly harbored doubts about the plane to come out of the closet. Even the *Washington Post* chimed in with an editorial asserting that the Lavi would only compound the strains on America's security assistance budget.[4] A political chain reaction was taking place.

The Lavi's supporters were anything but politically naive; they saw exactly what was going on. With opinion in Israel virtually deadlocked over the Lavi, the key to reversing the rising tide of negativity about the project was still to rally support in Washington. At the same time, IAI increasingly pinned its hopes on a teaming arrangement with an American company, most notably Grumman, with which the marketing talks continued to drag on. IAI actually was in a no-lose position. It stood to win the lion's share of work in Israel if any of our alternatives was adopted, and most particularly if Israel chose one of the F-16 alternatives. It therefore had every incentive to press ahead with the Lavi, in the hope that even if the program was terminated, by the time the decision was made the United States would have further sweetened the pot with additional work for IAI.

IAI and its allies therefore hit back harder and harder at the criticism that was being leveled at the plane and at the alternatives that we had proposed to it. Often the vitriol was personal: the Lavi's top test pilot, on flying the plane for a second time, generated headlines when he said, "I know nothing about politics; Zakheim knows nothing about planes." At other times it was a mix of substance and *ad hominem* put-downs: Menachem Eini dismissed our alternatives as "a bit fantastic," arguing that he and his team had examined all available alternatives before we had ever appeared on the scene. He went on to say that it was too early to tell whether "Zakheim had done a sloppy job," but that was because the Israelis still had to review the "thousands of pages" that we had left behind.[5]

There were also the purely substantive arguments. Dick Smull's team of contractors had met with the Israelis January 14–18, with GD meeting the Israelis on the 14th and 15th and McDonnell Douglas making its presentations during the remainder of that period. Seizing upon the fact that the visit had coincided with the Israeli air force's acceptance of its first F-16Cs, the Lavi's supporters argued that General Dynamics had made it clear that the cost of the F-16 would be much higher than I had advertised, because my estimates had not provided for the insertion of some systems slated for the Lavi. IAI spokesmen alleged that we had ignored the cost to the United States of terminating the program, which IAI estimated at about $500 million. They asserted that just as security assistance had risen in the past, Israel would find new ways to extract more money from the United States, either directly through security assistance, or through other, more subtle ways of working the budget books. They also contended that if Israel could not absorb three hundred Lavis the air force could not absorb three hundred F-16s either.[6]

There were, of course, rejoinders to IAI's rejoinders, as the arguments spun around and around. The termination costs were a pittance compared to the cost of continuing with the Lavi. Absent a massive infusion of new American funds, even finding another $100 million annually would not fund the project. An Israeli buy of as few as 150 F-16s increased unit costs by only about $2 million; a similar buy of Lavis increased that plane's unit cost by $13 million.[7] But the analytical game was over; whatever spin Eini or IAI or Arens put on the Lavi's costs, they could not overcome the consensus among independent Israeli analysts, our own analysts, and the GAO that the Lavi was simply a budget buster. By early February, the Israeli press revealed that Yitzchak Rabin himself had now also reached the conclusion that the Lavi was unaffordable.

During the first week of February 1987, Fred Iklé and I capitalized upon the flood of negative reports emanating from Israel as we made frequent trips to Capitol Hill to brief the key players on the Lavi. Congressman Lee Hamilton, who had finally received the GAO report on January 31 but had not yet released it to the public, made it clear to us that he was very unhappy with the program. David Obey, who chaired the Foreign Operations Subcommittee of the House Appropriations Committee, was even more outspoken. He would gladly have had all funding cut off, but knew he didn't have the votes to do so. And he was not going to take on his committee in a losing cause; he had too many other political fish to fry to make that sort of mistake. He did suggest that we brief several of his colleagues. Eventually I did so, but in the course of a hearing chaired by Lee Hamilton.

I met with Jack Kemp again on February 4. He twice apologized to me for hassling me about the Lavi; he remained committed to the plane but acknowledged that we had a much stronger case than many people had realized earlier. Jack Kemp is both an honorable man and an intellectually honest one. His reticence about the plane only six months after he starred at its coming-out party was for me the clearest signal yet that the political tide had turned in Washington. For once, the power of analysis—of what is called number crunching—had neutralized, though certainly not defeated, the political forces that were more natural to the Hill and from which the Lavi had drawn its strength.

Fred and I also visited Senator Dan Inouye, David Obey's counterpart as chairman of the Foreign Operations Appropriations Subcomittee, about two weeks after I had met Kemp. Like the New York congressman, Inouye was not about to change his views, but he was deeply disturbed by what we told him. He was particularly troubled when I said that most of the Israeli military opposed the plane. He turned to his aide, who confirmed that was indeed the case. I then added that Israel had to worry about one war at a time, and that its military readiness was suffering as a result of misplaced resources. Neither Inouye nor his aide replied to our points; they seemed saddened by them.

I also continued to keep in close touch with AIPAC, especially with Steve Rosen, a former analyst for the RAND Corporation who, since moving to Washington, had built up the lobbying organization's analytical capabilities. Steve knew that our numbers were not far off the mark. He did not try to persuade me otherwise; rather he simply wanted to keep tabs on how we saw the situation developing. For its part, AIPAC was quiescent. Its newsletter's report on our trip, which appeared in the "Heard on Capitol Hill" section, was devoid of all polemic and was rather dry.[8] That too was a sign of the changing political times.

There were, of course, some Jewish organizations whose support for the Lavi remained as strong as ever. Likud's American arm, the Herut organization, of which my dad had been vice president for many years, was one such group. Young Israel, a religiously Orthodox and politically Zionist synagogue movement, which included the synagogue to which I belonged, likewise asserted in its news-letter that "Israel's historical and technological need for the Lavi seems clear, regardless of cost."[9]

These organizations were vocal exceptions, however. More indic-ative was the release late in January of an analysis by the American Jewish Congress which put a rather different spin on the Lavi con-troversy. While somewhat sympathetic to the Israeli position, the AJC did not come down on the side of the Lavi. Rather it voiced in dis-creet terms what was becoming the overarching concern of American Jewish organizations: that the controversy not further embitter American-Israeli relations already reeling from the fallout of the Pollard affair. The AJC warned, "One thing is certain: this dispute could not occur at a worse time for both countries. Until it is resolved the continuing controversy can only have a seriously damaging effect upon the already overburdened, overly complicated relations between the two states."[10]

It was becoming increasingly clear in Washington that such a resolution would most likely result in the cancellation of the project. The case for saving the Lavi had been yet further complicated by another set of recent developments: the personnel changes in the Israeli embassy over the preceding year. To be sure, the ambassador, Meir Rosenne, was a strong Lavi supporter. But he was disliked by many in the administration for his hair-splitting legalistic approach to the Arab-Israeli conflict, and he did not have the personal clout on Capitol Hill that was the hallmark of Moshe Arens's tenure as ambassador to Washington. The senior military attaché, Amos Yaron, was a tank officer who recognized that the Lavi posed a threat to army programs he considered critical for Israel's long-term security. He was particularly close to Rich Armitage, who never missed an opportunity to dump on the project. Danny Halperin, the shrewd economics minister, had retired to private consulting in Israel. Though he gave pro-Lavi quotes at any and every opportunity, his presence was obviously no longer felt on the Hill. His replace-ment was a capable man but had few contacts in the capital. In any event, he was nowhere nearly as committed to the program as Halperin had been.

Nevertheless, while support for the Lavi in both Israel and Wash-ington was eroding, the political gridlock in Jerusalem meant that it would be very difficult actually to kill the program. There was no doubting the importance of Rabin's turnaround. Nevertheless, his opposition was not enough to move a National Unity government that was deeply divided on the issue, and whose inner cabinet included Lavi supporters Prime Minister Shamir, Foreign Minister Peres, Arens, Ariel Sharon, and the deputy Likud leader David Levy. Apart from Peres, the others took their cue from Arens, not Shamir. The prime minister's position on defense matters generally was rather ambiguous, and on the Lavi it was argued that he had "no particular position in the decision-making process."[11] On the other hand, Arens not only ensured Shamir's support, he also made it virtually impossible for his rivals for the Likud leadership, Ariel Sharon and David Levy, to be anything but strong supporters as well. To have done otherwise would have virtually guaranteed them calumny within Likud circles as traitors to national interest. Sharon in particular seemed to have forgotten that it was he who had originally ordered the General Dynamics study that now was the basis for the most attractive of the Lavi alternatives. In any event, Rabin was not prepared to terminate the Lavi on his own. He announced after our visit that the decision would be made by the entire cabinet, which was likely to support continuing with the Lavi.

Shamir was back in Washington in mid-February. He was amiable enough when we met at a reception in his honor at the Israeli embassy and called me his "friend" when he met with Cap Weinberger at the Pentagon on February 19. Only the day before, the secretary had publicly criticized the Lavi for the first time, in testimony before the House Foreign Affairs Committee, suggesting that it was a waste of American money for Israel to acquire an inferior airplane. The Weinberger-Shamir meeting focused more on Jonathan Pollard than on the Lavi, however. Cap had taken a very hard line against the spy. The Lavi issue came up briefly, and Cap indicated that he felt the GAO report supported our case (the report was finally going to be released on February 23), and that we would protect Israeli jobs even if the Lavi program was terminated.

Lee Hamilton held up releasing the GAO report until after Shamir left Washington. Its appearance triggered a new flood of newsprint about the Lavi. We had seen the report, of course; the published version included our lengthy comments on its contents. But for the public it came as a surprise, particularly given the spin that Hamilton put on it in the news release that accompanied its dissemination. Hamilton cited three major GAO findings. First, the GAO had "reconciled" the conflicting cost estimates to a level of $17–18 million. Second, it noted that there could be additional unforeseen costs (such as those that were resulting from the increase in Israeli wages). Most important of all, Hamilton quoted the GAO's finding that even the lowest estimate for the Lavi would exceed Israel's own spending cap and consume a huge proportion of America's annual security assistance to Israel.[12] The fact that Hamilton chose to state in his release that the report's "conclusions raise serious financial questions given projected annual costs" was an unequivocal signal to the Israelis that they could no longer look to Congress as the savior of the Lavi program.

We really had little to add to what Hamilton had already observed and to our comments that were included in the report itself. But to an insatiable press corps I also pointed out on the day after the report's release that the GAO's estimates, expressed as flyaway costs, omitted items such as production tooling, support, and spares. When incorporated into what is termed production unit costs, these factors resulted in a much wider gap between ourselves and the Israelis, one that the GAO analysis had in no way closed. I summed up by saying that we felt completely vindicated by the GAO's report.

Two days after Hamilton released the report, I made my case again in public, this time by testifying before his Europe and Near East Affairs Subcommittee with both Rich Armitage's deputy for Near East affairs, Ambassador Bob Pelletreau, and his State Department counterpart. It was the first and only time that the Lavi had ever been the subject of lengthy testimony on the Hill, and offered yet another signal that politically it no longer had any Teflon coating. Still, after nearly an hour went by before anyone mentioned the plane, I began to wonder if the subject would even come up. Pelletreau leaned over to me and whispered, "They're avoiding you like the plague," only to have Larry Smith, probably the most reflexive of Israel's supporters in the House, begin peppering me with questions and saying that I had a "personal obsession" with the issue. I ignored his remark, plowed ahead with my usual case, and encountered generally friendly questioning from most of the members of the committee who attended the hearing.

Hamilton himself never voiced direct opposition to the project. On the other hand, through careful questioning he enabled me to point out the dangers of going ahead with it. I stated unequivocally—and was not contradicted by anyone on the committee—that Israel could not meet its $550 million cap. For the first time I revealed our estimate that if only 150 aircraft were acquired by Israel, the cost of the Lavi would rise to $40 million per unit, or over $30 million even in terms of GAO's estimate. Mel Levine, a strong supporter of Israel, wanted my assurance that we would cover the termination liabilities for the aircraft. I was able to tell him that we would; that our estimate was that they would not exceed $400 million; and that higher estimates presupposed that everyone working on the project would be fired, which was simply not likely to be the case. Bob Toricelli, another strong supporter of Israel, emphasized his concern about the Lavi's impact on other defense programs, notably its tank warfare capability. I was also able to discuss the naval modernization program and to emphasize my support for what the Israeli navy hoped to accomplish. I left the Hill with the reinforced conviction that the battle in Washington was over; we had clearly neutralized our opposition. It was left for Rabin to lead the fight back in Israel.

Later that day, my father told me that he had met Prime Minister Shamir in New York. It was the first time they had seen each other in fifty-two years. The two men had hugged and reminisced about old times, two older men whose lives had been disrupted by war and misery and who had followed quite different trails from their village birthplace. Dad said Shamir had become quite maudlin and had tears in his eyes. The prime minister told Dad that his own father never made it to Treblinka; he was shot by a Lithuanian, as were his two sisters, who were killed by another Lithuanian, a merchant. Shamir had some nice things to tell my father about me and said that my heart was in the right place, but he added in Yiddish, *"Er macht fahr mir tzoris,"* meaning "He's giving me a hard time." No kidding.

By early March, Rabin's views were being widely reported in Israel and around the globe. They were, of course, known to the defense establishment somewhat earlier. Eini, recognizing that the end was in sight, had launched a final vicious attack on me on February 26. Calling me a *"kipah* [skullcap]-wearing religious Jew who claims to be motivated by love of Israel [but who] is in fact causing tremendous damage to Israel," he went on, quoting the Prophet Jeremiah, "It is about him that is has been said, 'Your destroyers and spoilers will issue forth from your own midst.'"[13]

Eini's remarks set off a storm in Israel and were carried in the American and European press. They were also counterproductive.

Everyone in the Pentagon and all whom I had worked with in the executive branch, even staffers on the Hill, called in their support. For the first time, I even received some encouragement from the Orthodox Jewish community, which had been conspicuous by its silence during the entire episode. Norman Lamm, president of Yeshiva University and a friend of two decades' standing, sent me a note bemoaning my treatment in the Israeli press. Louis Bernstein, who served several times as president of the Orthodox Rabbinical Council of America and whom I had known since I was a youngster, wrote openly about our friendship in an Orthodox Jewish magazine. It was all very heartening, since I had already informed Fred Iklé that I was planning to leave the department.

The support I received did not end with my American colleagues and friends. The religious press in Israel condemned Eini's statement as an "expression of Jewish anti-Semitism and as a particular insult to the many engineers and technicians working on the Lavi who wear *kippot.*"[14] I received calls from Israel, from people like Zvi Tropp, with whom I might have disagreed, but who had become good friends as well. And Yitzchak Rabin proved once again that he was a class act. On his orders, Eini was forced to make a public apology the following day, issuing a statement that said that his remarks were "out of place and should not have been made." Israel's defense attaché, General Amos Yaron, phoned my home on the weekend to let me know of Eini's retraction.

The apology was not enough for Rich Armitage. He told Cap Weinberger that he wanted to issue an official U.S. government démarche to the government of Israel protesting its behavior. He felt that my religion was entirely a private matter and certainly of no consequence to Israel. The Israelis had besmirched an American official and should not be allowed to get off the hook for what had been said. Weinberger agreed, and Rich called in General Yaron and reamed him. Amos, of course, was not to blame, and he dutifully passed DoD's message on to his superiors at Hakirya.

Although I was to leave the Pentagon on March 27, I hardly disengaged from my work until the last possible minute. Fred and I had agreed that I would hold back a formal announcement of my departure, so as to prevent the Israelis from seizing upon it as a sign that the U.S. government was somehow backing away from its opposition to the Lavi. The Israeli press circulated the "rumor" anyway.

In the meantime, reports kept emanating from Israel that the Israeli air force had pretty much rejected all of my alternatives in favor of a buy of more Peace Marble F-16s. The general staff formally recommended to Rabin that the Lavi be scrapped. Chief of Staff Levy told a Tel Aviv audience in late February that there were "ready-made, very attractive alternatives to the Lavi."[15] Levy was scheduled to depart the IDF in March to be replaced by none other than Dan Shomron, the IDF's most senior Lavi opponent, whose appointment by Rabin had been bitterly opposed by Moshe Arens. Arens had blocked Shomron's promotion to chief of staff four years earlier, and had appointed Levy instead.

If that were not enough of an indicator of where Rabin now stood on the issue, a flood of backgrounders emanating from his office made it clear that he too supported the air force approach, and would use the freed funds to move ahead with army and naval modernization.[16] An "Israeli source in Washington" told one of the local defense newsletters, "Things have gotten crazy over there. They have all divided up and they're fighting with each other."[17]

We had no problem with the current Israeli state of disarray on the issue. Nor did we anticipate having any difficulty with an Israeli decision to buy more F-16s. On the contrary, that was the alternative we preferred as well. An off-the-shelf buy of F-16s was the cheapest alternative and involved the least risk of schedule delays and cost increases caused by complicated integration of additional Israeli systems. We had formulated other approaches to satisfy IAI and the Israeli public that we were not simply trying to hamstring Israel's industry or the development of its high-technology sector. We were not wedded to any of them.

Weinberger was eager to have the project terminated once and for all. Late in February he refused to agree to a proposed State Department cable instructing Embassy Tel Aviv to indicate that the United States would permit a maximum of $550 million in offshore funds to be spent on the Lavi. "Kill it" was his attitude. Some of us thought that SecDef's tactics were still premature. General Yaron, who by now made it clear that he wanted the Lavi to go away, advised me that Weinberger's attack on the Lavi when he testified on the Hill had hurt our cause. I felt that Yaron was right, and, together with Armitage, tried to persuade the secretary to agree to moderate the tone of his opposition to the program, at least for a bit longer. Weinberger held his ground and refused to concur to even a modified form of the State cable that still implied American acquiescence to the project's continuity.

Early in March I traveled up to New York to brief the Presidents' Conference, the umbrella organization of American Jewry, on the Lavi. The briefing was relatively uneventful, except for a rude interjection by the IAI representative in New York, who had somehow made his way into the gathering. Chatting with some of the invitees at the end of the meeting, I was told by Israel Singer, the executive director of the World Jewish Congress, that my father had been viciously attacked over the Lavi at a WJC meeting, which he had attended as a former vice president of its American section. Singer said that he had just returned from Israel and was reeling over Rabin's decision to oppose the plane. "Maybe your dad can come again to meetings," he told me. It was the first I had heard of the public brutalization of my father; he never mentioned that or any other incident to me. At the time he was seventy-seven years old.

The Israelis had promised us an evaluation of our alternatives study; on March 9, Dick Smull received a sparse three-page document from the New York mission. Whatever the press might be reporting about Rabin and the air force, the "preliminary" MoD reaction to our alternatives made it clear that formally the Israeli government was sticking to its position that nothing had changed. The summary stated that the Israeli air force felt that none of the alternatives were an "adequate solution" to its present operational requirements; in particular, none could substitute for the Lavi's planned role as an advanced training aircraft. Moreover, once the alternatives were adjusted to account for all of the Lavi's technical and operational specifications, the cost of each was higher than that of the Israeli plane. The Israelis also reiterated that our alternatives had provided for no termination costs, nor had they accounted for Israel's lower labor rates.

There was little that was surprising in the Israeli response, which was unclassified, and, we assumed, primed for public consumption. The case for the Lavi as a trainer was a somewhat new twist on the IAI theme, however. Whether it was worth Israel's while to beggar its defense budget in order to acquire a training aircraft was an issue that the summary naturally did not address. We concluded that the summary meant nothing, and ignored it virtually from the day we received it.

As the day of my departure neared, the Lavi's supporters were shocked by another unfavorable report, this time by Israel's auditor general, that blasted the project and all who were associated with it. The auditor general found that the cost of developing the Lavi had far outstripped development costs of equivalent aircraft. He criticized IAI for deliberately understating the plane's costs. And he castigated the ministry of defense, the Israeli air force, and Amos Lapidot in particular for not undertaking any serious analysis of the implications of the Lavi for other air force projects. The auditor concluded that if the project was continued, Israel would have to terminate virtually every other one of its major weapons programs. The report was meant to be secret, but it was about as secret as most political documents are in Israel; that is to say, it leaked all over the Israeli press and soon found its way into the American media as well.

In the meantime, news of my leaving had made its rounds in Israel, and the press was full of speculation as to why I was moving on. The answer was simple enough: I had managed to put away virtually nothing in savings on my government salary, and had been offered a much better paying and quite interesting job at System Planning Corporation, a Virginia high-technology consulting firm. I also knew that I would continue to advise the Defense Department, particularly on the Lavi, though fortunately that prospective role did not become known to the Israelis for some time. When they did find out, they were none too thrilled. But by that time, it was clear to everyone that the issue was no longer whether the Lavi would be terminated, but when.

17

The Lavi Is Grounded

I no longer worked for the government, at least on a full-time basis, and therefore was no longer in charge of coordinating policy on the Lavi. That task formally devolved to my successor, Dennis Kloske, who had been Deputy Secretary Will Taft's assistant. In practice, however, the real office in charge was Rich Armitage's International Security Affairs, though its work consisted primarily of monitoring developments in Israel and preparing talking points for senior officials who met with Israeli government representatives.

Despite, and perhaps because of, my departure, the press, especially the Israeli media, seemed eager to hear what I had to say about the project. Since I no longer had to follow the official public affairs line, many reporters figured that I would be much more frank about my own motivations, as well as those of my erstwhile colleagues. I gave a slew of interviews immediately prior to and shortly after my departure. The questions, and the answers, were boringly repetitive. Yes, I really did believe that the Lavi was a burden on both Israel and the U.S. taxpayer. No, I did not regret the role I had played. Yes, I was proud of my effort. No, I did not see a conflict between my activities at DoD and my Judaism.

As I made my move out of the Pentagon, the Lavi issue appeared stalemated. It continued to remain intertwined with the Pollard affair: in mid-March, Pollard and his wife had been sentenced in federal court. The presiding judge, in receipt of a secret memorandum from Cap Weinberger, and showing no sympathy whatsoever for the defendants, sentenced Pollard to life imprisonment while his wife received two concurrent five-year jail terms. The Knesset had voted overwhelmingly to seek Pollard's release, a move supported by 90 percent of the Israeli public.

At the end of March, the Israeli colonel who had masterminded Pollard's activities in Washington resigned from the air force. He had caused a stir because immediately upon Pollard's detention by the FBI, he had returned to Israel for reassignment to the command of Tel Nof Air Base. Tel Nof happened to be where American F-16s usually landed in Israel, and an outraged DoD had made it clear to the Israelis that it resented the colonel's appointment and that there would be no cooperation between the two air forces at Tel Nof while he remained in charge.

Israeli journalists peppered me with questions about Jonathan Pollard's sentencing. I had not changed my views about the man, and was quoted in *Yediot Achronot* as saying that if Pollard felt he needed to help Israel he should have made *aliyah*. My remarks infuriated his many supporters; no doubt they also added fuel to the Lavi fires that IAI continued to stoke.

There were other developments that were complicating the case for the Lavi. Israeli arms sales to South Africa were a constant irritant in American-Israeli relations, and the administration was pressing Jerusalem to terminate its long-standing and lucrative relationship with Pretoria. At about the same time as Pollard's former spymaster Colonel Sella was making headlines again, Rabin announced that he could not anticipate a cabinet decision on the Lavi until after the U.S. Congress received a presidential report on South Africa in May. Rabin reportedly felt that "Israel can either cancel its military ties with Pretoria or drop the Lavi, but not both, if it does not want to ruin its aircraft industry."[1]

Finally, Rabin had to consider the increasing and unambiguous threats by Lavi workers to throw their electoral weight against any party that dumped their project. Israel Aircraft Industries' twenty thousand workers and their families represented enough voting power to control at least five Knesset seats. That was enough to tilt the delicate balance of power between the Likud and Labor alignments. With the unions and IAI both vocally supporting the Lavi, it would take a brave minister indeed to torpedo the project. Only if a coalition of votes on the part of the two parties could somehow be cobbled together in cabinet could the Lavi be dropped without political consequences in the Knesset. But Arens was still holding the line on the Likud; Rabin had so far failed to forge a coalition. It was not at all clear that he would be any more successful in May than he was in March, and public criticism of his delaying tactics was getting sharper.[2]

On the positive side of the ledger, it was widely reported that the American Jewish leadership had refused to support the creation of a special Lavi fund. Shortly after I had addressed them, the Presidents' Conference leaders had visited Israel and had paid a visit to IAI. It was there that the subject of an investment fund for the Lavi was broached indirectly, and was completely ignored. Privately, several Jewish leaders were unequivocal about their opposition to the project. The story was carried in the Hebrew press, which speculated that IAI would not have aired the idea without first coordinating with the MoD.[3] The rebuff sent an important signal to all sides of the debate that the American Jewish community could not be counted on to affect its outcome.

Another report that gave me immense pleasure was the story of my supportive role in the Israeli submarine program that appeared in the American-Israeli paper *Yisrael Shelanu*. The American public, and certainly the Israelis living in the United States, of which there are hundreds of thousands, had only known me as the evil demon behind the effort to kill the Lavi. It was nice to be viewed in a completely different light.

Rabin, meanwhile, was receiving more bad news about the Lavi's costs. Late in April, a defense ministry study team concluded that the purchase of 150 Lavis would cost between 46 and 56 percent more than the equivalent buy of F-16s from the United States.[4] That conclusion was not much different from the General Dynamics analysis some months earlier.

Rabin still had not found a way to build a nonpartisan anti-Lavi coalition. But Finance Minister Moshe Nissim finally afforded him the opportunity to do so. At an April 21 cabinet meeting, Rabin again raised the issue of an infusion of additional funds for the Lavi. Nissim vehemently opposed the idea. In response, Rabin asserted that only such an infusion could save the plane. The project was already on a very short leash: the new budget provided only six months' funding for continuation of the Lavi. By demanding a special fund, however,

Rabin was forcing Nissim into a corner, and was setting the stage for the Likud finance minister's formal defection from the pro-Lavi camp.

Within a month, Nissim hardened his position. On a May 21 tour of IAI, just after the defense staff formally briefed a cabinet session on the steady rise in Lavi costs, the finance minister reiterated that there be no special Lavi fund. In fact, Nissim went even further. He stated that continuation of the program would call for reductions in the budgets of all the other ministries, and of the defense ministry first and foremost among them. By responding to Rabin in this manner, Nissim was attempting to draw other Likud ministers into opposing the project. It was easy enough for Arens to push it; he led no ministry and had no budget to defend. Nissim was telling his other Likud colleagues in the cabinet that not only would Rabin's budgetary ox be gored, but theirs would as well. No message conveys more impact to a minister than the threat of a budget cut. At a minimum, Nissim had guaranteed that support for the plane, even from Likud ministers, would henceforth be considerably more muted.

Late in May 1987, word leaked that the inner cabinet would appoint a special six-minister task force to formulate a recommendation to the government on the future of the Lavi. The appointments were expected to be made at the cabinet's regular Sunday meeting that was to take place on June 1. The atmosphere leading up to the meeting was downright poisonous. A few days earlier, IAI's acerbic president, Moshe Keret, had blasted Avihu Bin Nun, the IDF's planner, for supposedly campaigning against the Lavi. IDF "sources" responded by rejecting Keret's criticism. At about the same time, and shortly after IDF representatives told the Knesset that they supported the U.S. position on the Lavi, a left-wing member of the Knesset finance committee named Dedi Zucker asserted that IAI had deliberately distorted the plane's costs. He claimed that IAI had based its estimates on a buy of 450 aircraft, which he argued was five times greater than the Israeli air force's requirement.

There was more. Just before the cabinet meeting, the Knesset state control committee reversed its earlier decision not to publish the auditor general's report and decided that it would only apply some national-security-related deletions before releasing the document. On the other hand, the Israeli press reported that workers in plants engaged in Lavi-related development were organizing to apply pressure to the cabinet ministers. And the heads of the enterprises involved in the project met to rebut an anti-Lavi statement made by the board of Tadiran, a major Israeli aerospace company that was not a Lavi contractor.

Apprehension about the program's fate had spread to the Lavi's 730-odd American subcontractors as well, though none was prepared to agitate the Congress to save the program. The Long Island paper *Newsday* made headlines in Israel when it revealed that Grumman seemed willing to increase its participation in the Lavi.[5] But when a Grumman team visited Israel in May to conduct still more talks, it failed to reach agreement with IAI.

The cabinet devoted about five hours to its June 1 discussion of the Lavi. Chief of Staff Dan Shomron, Director General Ivri, Zvi Tropp, Menachem Eini, Moshe Keret, and Mordechai Hod of IAI briefed the meeting. The ministers were shocked by some of the material that the MoD provided them. It was reported that the original 1980 Lavi decision was taken with no evaluation of its economic impact or its effect on other defense programs. Data about the plane's costs were still unclear; details of alternatives had yet to be formulated. While the IAI representatives lobbied for the plane, Rabin was circumspect, although he let David Ivri speak out in its favor. Rabin instead returned to his demand for more money for the Lavi; it was not a defense project but a national project, he asserted, and required national funds. Nissim held his ground, and no additional monies were granted.

In a separate interview, Dan Shomron once again weighed in against the project. With no one muzzling his views, the new chief of staff warned that completion of the Lavi not only would hurt other military procurement, but would affect training as well. In a country where military training and readiness was critical to national survival, Shomron's statement was yet another shattering blow to the beleaguered program.

Though it took no action on finances at its June 1 meeting, the cabinet did appoint the Lavi task force. The team was composed of Labor ministers Rabin, Peres, and Gad Yaacobi, the minister of economics who had recently spoken out against the plane. Arens, Sharon, and Nissim represented the Likud faction. Joe Bavaria, our attaché in Tel Aviv, told me over the phone that the task force was a ploy to cover up for the "kill." He said that Peres had now joined Rabin in opposing the project. If Nissim could be induced to oppose the plane, the task force's recommendation would be negative, and termination was likely.

I was also hearing that Rabin was quietly encouraging opposition among Israeli industrialists by pointing out the risk that the Lavi posed to all other military projects. Several Israeli industrialists told me at a conference I attended in early June that engineers were having trouble finding jobs in Israel, and that both Tadiran—which already had publicly attacked the project—and Rafael, the government's own research and development corporation, were being hurt financially by the diversion of resources to the Lavi.

Whatever the grumbling of its competitors, IAI continued to market the Lavi as heavily as it ever had. In anticipation of the Paris Air Show, IAI took out an ad on the back page of *Defense News* that stressed its joint American-Israeli origins ("getting there together...").[6] The plane figured prominently at the company's booth at the Paris show.

Rabin was now scheduled to visit Washington in late June, and it appeared that no decision on the Lavi would be taken until after his visit. The cabinet deferred a vote at its June 21 meeting and simply announced that it would review the matter during its next several weekly sessions. Three ministers had spoken against the plane, five in its favor. Among those most vociferously for the project was Housing Minister David Levy, one of Arens's most bitter rivals in the struggle to succeed Shamir as Likud's leader. Levy, a former worker, prided himself on his ties to the working class. He was not about to let anyone outdo him in his support for jobs, and especially not the professorial Arens. Rabin, having just returned from the Paris Air Show, stated that there was a glut of aircraft in a shrinking international market. He reportedly told the meeting that there was no basis for anticipating foreign sales of the plane. With no budget increase there would be no Lavi.[7]

It remained unclear whether Rabin would wait until the program died under the weight of public opposition or would actively precipitate its demise. When I chatted with Abba Eban after a speech he had given to a Washington think tank the day after the cabinet meeting, the former Labor foreign minister told me that the government was trying to generate a Knesset vote on the Lavi. Eban said that the chamber's foreign affairs and defense committee had reported against the project. He believed that a majority of the Knesset, like the military, opposed the plane. But he felt that the decision was best taken by Rabin and his cabinet colleagues.

Reports emanating from Israel indicated that the divisions within MoD and the military over the project were leaving ministers thoroughly confused. Rabin decided that he would use the occasion of his visit to Washington to obtain clarification regarding the administration's willingness to provide financial support for the project's termination. At issue was the cost of the termination liabilities, for which I had much earlier received Weinberger's approval, as well as an ironclad American commitment to a long-term offshore funding program for projects other than the Lavi. Initially, both Weinberger and Shultz had been reluctant to meet with Rabin until after Israel had decided upon cancellation, but they both reversed their positions. Rabin's public explanation for the trip was that he was seeking a "tentative" idea of America's response to a decision either to cancel or proceed with the Lavi. The response would serve as input to the cabinet's deliberations.

I was still seeing Fred Iklé and Rich Armitage frequently, and pressed them to accept the Israeli demands and to tell Rabin as much when they saw him. It was the only way to take the program out of its misery. They both agreed, as did Weinberger and Shultz. Whatever their differences in other matters—and the Iran-contra scandal was now at its height—the prevailing view inside the administration was that everything should be done to facilitate any effort by Rabin to terminate the program.

Rabin's visit began on June 29, 1987, and coincided with the public release of the Israeli auditor general's report on the Lavi. It ran to nearly thirteen hundred pages and was accompanied by a special forty-page report on the Lavi that the auditor general released at a press conference. Described as "the most critical in Israel's history,"[8] the report confirmed the U.S. administration's position on the cost issue. The governor of the Bank of Israel, Dr. Michael Bruno, also weighed in, arguing that there was no economic justification for the program and that the government "had no choice but to halt the project immediately."[9]

The Israeli government issued a point-by-point rebuttal of the report, while Arens continued to assert that nothing had changed and that he expected increased levels of American military assistance to Israel. But lasting damage had been done to the program. Both the Israeli auditor general's report and Bank of Israel governor

Bruno's opposition undermined much of the residual support that the program commanded on Capitol Hill. And Tom Pickering told the Israeli press that while the United States would permit its security assistance funds to be used for payment of termination liabilities, the level of assistance would not be increased.

Upon learning that the report had been released, I phoned Amnon Neubach, still associated with Shimon Peres and now at the foreign ministry, to get a sense of where the cabinet was heading. Neubach asserted that Peres, Rabin, Shamir, and Finance Minister Nissim had agreed to press for maximum American financial support in exchange for the program's termination. Neubach claimed that Peres had originated the plan. He added that IAI supposedly was seeking $1.4 billion in termination liabilities, or about three times what we estimated those liabilities would cost.

On the day Rabin arrived in Washington, I received a call to come to his suite at the Grand Hotel the following day. The next day the meeting was postponed to the thirtieth. I was not entirely sure why he wanted to see me. Several weeks before, I had happened to fly up to New York on the same shuttle as defense attaché Amos Yaron. "Look," he said, "everybody in Israel thinks you're consulting on the Lavi, so why not just admit it?" But I did nothing of the kind; Fred Iklé and Rich Armitage had preferred that my role remain strictly off-the-record.

I arrived at the hotel at about the time Rabin's own entourage swept into the place. I let the group take the elevator up first, and followed a few moments later. I was admitted to suite 706 and waited about five minutes in the outer room before being asked to go in and see Rabin.

Rabin was alone in the sitting room. He asked me to sit and asked how I was. He had never been so obviously solicitous before. He said he had met with Weinberger (I had already been briefed by Weinberger's military assistant) and then went on to say that he was angry at what Eini had said about me. Eini was wrong and he had chastised him. I told Rabin that I knew what he had done and I thanked him for it. Rabin said that he would not call Eini into the room unless I agreed. Eini would apologize to me personally, but Rabin again emphasized that he would not call Eini in if I did not want him to. He then said that he, Rabin, very much wanted Eini to come in and for a third time said it was up to me.

I said to Rabin, in Hebrew, "What happened, happened. That is behind us." Rabin answered me in Hebrew, "I want to thank you for all that you did." Then for a fourth time, he said Eini would enter only if I agreed. I said fine, and Rabin called his people in—Eini, Abraham Ben-Joseph, Yaron, and others in his entourage. Eini sat on the sofa with Rabin; there was some space between them. Everyone else took a chair. Rabin looked to Eini; the Lavi program manager did what he had to do. I felt a bit uncomfortable about it all, but I accepted the apology, and said that it was behind us. The tension in the room evaporated immediately.

Rabin reviewed his meeting with Weinberger. He had told the secretary how difficult the Lavi problem was for Israel. There were three alternatives: to proceed with the Lavi, go with another aircraft, or simply kill the Lavi. Only the latter two alternatives were viable, Rabin had said. He added that Weinberger had reminded him that he had always thought the Lavi was bad for Israel.

Rabin then said he had rejected three of my five major alternatives. Only the F-16 alternatives were viable. That is, Israel could either acquire some more planes directly from the United States, or jointly develop a new Israeli version of the aircraft. Rabin needed more fighters by 1992, however, and an off-the-shelf purchase was the best way to meet his requirement. Weinberger had agreed to approve the sales.

Rabin also needed $300 million in continued offshore procurement funding; Weinberger had agreed. He needed the United States to fund the Lavi's termination liabilities; Weinberger had agreed to that too. Rabin said that there were problems with some of his other requests, such as to maintain the directed offset program, or to add $100 million in offshore funds for the next two years. He also worried about certain of the nonrecurring costs that would be associated with a buy of American aircraft. I told Rabin that I thought Cap would be tougher on extending the offsets than on offshore increases. Rabin agreed. We then talked about the prospects for congressional approval of these arrangements; I felt that in the current environment the Congress could not be taken for granted. Ben-Joseph said that the administration, not the Congress, was opposing the plan.

I asked if Rabin was alone in formulating these proposals. He answered, "I initiated my trip." I said that the ball was in Rabin's court. Weinberger needed a list of all of Israel's proposals, even if some were not acceptable. Rabin acknowledged that he had to furnish such a list. I said that the ministers had to "scope" the problem and needed to be aware of what was at stake. I suggested that MoD immediately obtain accurate termination liability costs. The list should also include the proposed new program for cooperative research and development of anti–tactical ballistic missiles as well as Israel's request for relief from nonrecurring costs associated with a new aircraft purchase. As I later told General Yaron, the purpose of the list was to indicate as clearly as possible the many important Israeli requests that would not be filled if Israel proceeded with the Lavi.

I was planning to visit Israel for the Sukkot holiday in September and told Rabin that I hoped the trip would not generate any publicity. Rabin replied, "Don't be so sure. You pay a price as a public figure." We then talked about how he could manage the stamina for such a short trip to the States. He said he took half a sleeping pill, slept soundly, and didn't get jet lag. I mentioned Michael Heseltine's thirty-six-hour visit to the States to obtain funding for a British SDI program, and said that he had worn out his staff. Rabin seemed rather unimpressed by the British minister of defense, whom Margaret Thatcher had fired some time after that visit but who later resurfaced in Prime Minister John Major's cabinet.

I left Rabin's hotel after about an hour. It was the longest private conversation I had ever had with him. He had been exceedingly gracious, and need not have put Eini on the spot. I reported the entire meeting to Weinberger's military aide, who in turn briefed the secretary.

Rabin then outdid himself the next day. Speaking before a packed gathering at the Washington Institute for Near East Policy, he surprised me, and I suspect everyone else who was there, by thanking me for the work that led to Israel's reevaluation of the Lavi project. After his talk I was besieged by the press, this time to comment on my vindication. I reported the event to Rich Armitage, who had already been briefed on my meeting the previous day. Rich reiterated that I should continue to report to him and to Weinberger on the Lavi as long as it remained a major issue. Rich said that Cap had asked him if I was still solid on the Lavi and Rich had replied, "Second only to you, Mr. Secretary."

Rabin returned to Israel to announce his opposition to the program and to brief a cabinet that was divided virtually up the middle on the Lavi issue. One senior Israeli official told the *Jerusalem Post* that Israel could not afford to have a decision on a matter of such national importance be decided by a 13–12 vote.[10] Shamir and Peres were reportedly seeking some compromise to keep the plane alive, but Rabin offered no hope from Washington. He had found that even Israel's friends on the Hill were advising that the project be terminated.

The following week, Weinberger asked to see me, and we discussed my conversation with Rabin. The secretary thought that Rabin was very perceptive on the question of offsets; there would be real difficulty extending them for several more years. Weinberger then said he worried about Peres, who had seen the Lavi as the embodiment of Israel's technological advances. He asked if we should offer more to the Israelis than we already had. I advised him not to. The package was good enough to convince the minister of defense; it should be good enough for the rest of the cabinet.

Meanwhile the debate in Israel seemed interminable. Week after week the cabinet discussed the issue with no resolution one way or the other. The Pentagon began to release broad hints of what programs would be jeopardized if the Lavi project was not discontinued. A report in *Inside the Pentagon* listed not only the ATBM project that I had discussed with Rabin, but also cooperative research on reactive armor for tanks and warning systems for aircraft. Yaron asked to see me to complain about the report. I had actually spoken to the author of the report beforehand and had told her that the Pentagon might slow down the projects, but would not cancel them. Yaron said that reports of Pentagon retaliation only stiffened the backs of the Lavi supporters. I suspected that those were the very people who were behind the stories. They had precisely that objective in mind. Yaron told me that he had sent a list of vulnerable projects to Rabin, as I had suggested. He also said he thought Peres and Shamir would ultimately vote against the plane.

In fact, there was still very little movement in the cabinet on the issue. Moshe Nissim announced that he would vote against the project. Shamir responded by saying that he still supported it. There were still not enough votes to scuttle the plane. It looked as if at least a month would pass before the cabinet faced up to the crucial vote.

In the meantime, Abba Eban's prediction about an impending Knesset vote proved accurate. The foreign affairs/defense and finance committees decided to hold joint hearings on the Lavi before putting the matter to a vote. Late in July they invited Arens, Sharon, and Weitzmann, the three former defense ministers serving in the government, to make their respective cases for and against the plane. Sharon and Arens briefed the committees jointly; Ezer Weitzmann chose to make his case on his own.

Arens, still refusing to come to terms with budgetary reality, argued that Israel had thus far not spent a dollar on the plane and, "if it managed wisely," it would not have to spend a dollar on it in the future either. His message resonated with some but not all of the members, including members of his own party. Meir Shitrit of Likud summed up the views of the skeptics, however, when he stated that he could not support the project if the entire general staff opposed it.[11]

Rabin also addressed the joint committee. Meeting with the members on the day of their vote, he stated that the Lavi represented no material increase to the IDF's capabilities and that he would not be comfortable sending the Lavi against the MiG-23 or MiG-29, should they be fielded by the Syrians.[12] Still, Arens prevailed for one last time. The two committees voted overwhelmingly on August 9 to support continuation of the program. Twenty members supported the Lavi, and only three opposed while three abstained. The Knesset resolution, which was nonbinding, was viewed as an important political statement; few Knesset members were as yet willing to face the wrath of the IAI unions.

One day later, the United States responded to the Knesset vote. For the first time, the State Department bluntly urged Israel to terminate the Lavi. Shultz's spokesman made it clear that Israel could not afford the plane, even on the basis of its own cost estimates, "without crowding out other projects."[13] Shultz himself sent a personal message to Shamir and his senior ministers in the Israeli cabinet urging them to drop the project. He argued that continuing with the Lavi was against Israel's own interests, since there were cheaper alternatives. He also reaffirmed his promise to Rabin that the United States would help Israel through its transition from the Lavi to an alternative program.

The State Department announcement and Shultz's note proved to be the pivotal events that ultimately forced the cabinet to a vote. As Shultz later recalled in his memoirs, he had changed his mind about the program and was now convinced it was a mistake.[14] He was determined to support Rabin and the plane's opponents in the cabinet and to bring about a quick resolution of the matter. It was assumed in Washington that there was no way that the government of Israel would defy its best friend in Washington on what was fundamentally an economic issue. The cabinet might delay a bit longer, but a vote was now inevitable.

Shultz's letter was timed to affect the cabinet meeting scheduled for August 16. Armitage had asked me to provide a memorandum on the implications of cancellation for the U.S. Security Assistance Program, which Weinberger used when he met on August 13 with Moshe Arad, the new Israeli ambassador. I learned afterward that Cap had given Arad a very hard time on the Lavi issue, telling him that it was a waste of taxpayers' money.

The Lavi's supporters were not about to concede defeat. They were fazed neither by Shultz's pressure nor by the rumor that Nissim had threatened to resign if the Lavi was not cancelled.[15] Shamir responded to the Shultz note by stating that he would propose that the Lavi be treated as a national project with all ministries contributing to its preservation. As a kind of compromise, presumably with the intention of satisfying his Likud colleague Nissim, the prime minister advocated a limited production run of seventy-five aircraft, which would have driven up the plane's unit cost though not the total cost of the program.

Apparently, Shamir had received assurances from Larry Smith, Mel Levine, and Charlie Wilson that if Israel decided to keep the project alive, the Congress would prevent the Pentagon from penalizing Jerusalem for its decision. For their part, Rabin and Nissim were now committed to asking the cabinet for a vote to cancel the project. There was also talk of a possible compromise, with development continuing to be funded but no further plans for production.

The press was by now overwhelmingly in favor of cancellation. Some writers bemoaned the fact that the Israeli government had only come to grips with the issue because the latest American statement sounded like a "diktat." Others noted that State was doing the cabinet a favor, by forcing an end to the logjam and taking the ministers out of their political misery. The margin was predicted to be razor-thin; *Yediot Achronot* headlined that there were ten votes for, ten against, with four more still on the fence.[16]

Shamir's proposal did not fly any more than the Lavi did. There was little enthusiasm for funding the plane out of ministries other than defense. Nor was there much support for Peres's compromise that would have provided the plane with interim funding until the matter was finally resolved. On the other hand, there also was little enthusiasm for a formal vote. For the seventh time, and after six hours of debate, the cabinet put off the vote, this time for another two weeks, ostensibly in order to continue its search for a politically safe consensus.

The ministers had actually taken a straw vote at the meeting, and could not even agree on its results. Peres claimed that there were twelve ministers for and twelve against; Rabin claimed a vote would have gone 13–11 to kill the project. There was also some speculation that Rabin agreed to the postponement because he did not dare risk a tie, which would have permitted the project to survive.[17] On the other hand, we heard via the embassy that the cabinet would have voted 13–10 to scrap the project, with Modai, now a minister without portfolio, and Yigal Hurevitz, who likewise did not manage a ministry, both abstaining.

The postponement gave IAI, Arens, and Co. one more reprieve to head off the inevitable. It also increased the pressure on the four fence-sitters—Modai; Hurevitz; Yaakov Tzur, the absorption minister; and Rabbi Yitzchak Peretz, the Minister for Religious Affairs who represented the Sephardi religious Shas Party. And there was one more possible waverer—Shimon Peres, who for some time had been rumored to support Rabin and who clearly was disappointed when his own compromise proposal had failed in the cabinet.

Peres was the first to commit to a switch. Shortly before the vote scheduled for August 30, the foreign minister formally called for the cabinet to consider alternatives to the Lavi. He suggested a Lavi 2000, which would involve Israeli participation in the F-16 and the new American advanced tactical fighter. It now appeared that with Peres against the plane, the opponents had the majority. But that seemed to depend on Nissim, Peretz, and two other ministers from the smaller parties. Labor could at best muster its own ten cabinet votes against the plane, but that left it three short. Health Minister Shoshanna Arbelli-Almozlino, a Labor member, had long been a strong Lavi supporter, and she could not be counted on to switch her vote. Nissim was the key, and he was under tremendous pressure from his Likud colleagues not to vote against them. On Saturday night, August 29, Tom Pickering delivered yet another message to Shamir, Peres, and Rabin, urging the cabinet to terminate the program and again offering U.S. assistance, including an increase of the offshore program to $400 million.

The following morning, I received a message at home to call

Chaim Shaked, the Israeli naval attaché. I reached him just after noontime. Chaim came quickly to the point. The cabinet had finally come to grips with the Lavi issue. The vote was 12–11 to cancel the plane. Four ministers had abstained, but the critical abstention was Arbelli. She finally had given in to tremendous pressure from the Labor leadership and did not vote for the Lavi. Nissim voted against the plane, the only Likud minister to do so.

THE LAVI VOTE: KEY MINISTERS

For	Against	Abstaining
Arens *(Likud)*	Peres *(Labor)*	Arbelli-Almozlino
Shamir *(Likud)*	Rabin *(Labor)*	*(Labor)*
Sharon *(Likud)*	Nissim *(Likud)*	
Levy *(Likud)*	Weitzmann *(Labor)*	
Modai *(Likud)*	Yaacobi *(Labor)*	
Sharir *(Likud)*	Navon *(Labor)*	
	Bar Lev *(Labor)*	
	Tzur *(Labor)*	

The cabinet adopted a compromise proposal by Peres to allocate $100 million to Lavi technologies; it was Peres who got Arbelli to change her vote and support his compromise. "Immediately after the vote Mrs. Arbelli-Almozlino ran from the government meeting and burst into tears."[18] At long last, the Lavi was grounded.

18

Aftermath

The Lavi decision caused an uproar in Israel. There were massive demonstrations orchestrated by the IAI unions. Workers blocked the main roadways in Tel Aviv and Jerusalem, as well as Highway 1 that links the two cities, by creating barricades of burning tires. Workers' committees pleaded with Prime Minister Shamir to freeze the decision for three months while studying alternatives that might save the program. Shamir would not commit himself.

For three days the workers continued to demonstrate, snarling traffic on Israel's main highways and delaying flights at Ben Gurion Airport. Demonstrators marched to Labor Party headquarters in Tel Aviv, venting their spleen at Peres and what they perceived as his betrayal of the program. There were also demonstrations outside Shamir's home in Jerusalem, which was guarded by riot police on horseback. Workers also demonstrated outside the finance ministry, and at the Wailing Wall, where they blew shofars to mourn the decision. Some demonstrators even tried to bring a model of the Lavi to the Wailing Wall plaza, but they could not get by the police.

IAI and the unions announced that the decision would touch off a brain drain. Within days of the cabinet's decision, hundreds of engineers had lined the streets in front of the American and Canadian embassies seeking immigration visas. But the cabinet decision was not reversed. Shamir announced that "we must not teach the public that government decisions will not be kept."

The U.S. government hurriedly began to make good on its various promises. It agreed to sell Israel off-the-shelf F-16s, with Israeli components, in effect an updated Peace Marble II. The Israelis ultimately purchased seventy-five aircraft, though at a price considerably higher than they had hoped to pay. Still, as even David Ivri pointed out, it was about $5 billion lower than what the Lavi would have cost them.[1]

Ivri quickly negotiated the additional $100 million in offshore funding that the United States had promised and obtained an extension of the directed offsets agreement beyond fiscal year 1988, when they had originally been meant to be phased out. The two countries also began a complicated negotiation on the size of the termination liabilities; virtually all the American companies supporting the Lavi took the news in stride.

It took somewhat longer for the United States to agree to fund the lion's share of an Israeli ATBM program, later called the Arrow, along the lines that I had recommended in my final briefing to Rabin and the MoD. The Israelis originally hoped that the United States would foot 90 percent of the bill; the initial U.S. position was to offer 50 percent. I felt that 50 percent was too low, and urged Assistant Secretary Richard Armitage to raise the offer to 80 percent. The two sides ultimately agreed on 80 percent, and much of the work went to IAI.

The United States continues to fund the Arrow, currently accounting for about 70 percent of the cost of the second phase of the missile program. The Israeli government has yet to commit itself to producing the Arrow, however. The program's costs have continued to rise, leading some analysts to evoke the memory of the Lavi.

The naval modernization program also moved ahead. It was not until the Gulf War that Israel finally solved its submarine offshore procurement problem, however, when Germany agreed to help finance the sale of its boats to Israel. On the other hand, the Saar V corvette program moved along smoothly. The contract was awarded to the Ingalls yard in Pascagoula, and the first ships were launched in the early 1990s.

Finally, within months of the cabinet decision, the United States and Israel signed a ten-year memorandum of understanding (MOU) that brought many of the free-trade elements of the U.S.–Israeli commercial sector to their military commerce. It was a major stride for Israel, opening up new and large markets to its domestic indus-try. Although the negotiations had begun under Cap Weinberger, he was not in the Pentagon to reap the credit for its conclusion; Frank Carlucci, his onetime deputy, was now secretary and signed the MOU with Rabin.

The Lavi story was, if anything, a case study in the depth and complexity of Israeli-American relations. It was also, however, a vivid illustration of the degree to which people operating at below the highest levels of government could exert tremendous influence on decisions affecting hundreds of millions or even billions of dol-lars. Although the Israelis inaccurately cast me as the evil spirit behind the attack on the Lavi, they were correct in their assessment that, despite my relatively low rank as a fourth-level administration official, I exerted influence on the fate of the program. Congressional staffers, such as Jim Bond of the Appropriations Committee, like-wise were able to exert far more influence over this and other pro-grams than their formal government rank might have implied.

Such developments should come as no surprise, however. As the size, complexity, and workload of the government have grown, often exponentially, the ability of the topmost decision-makers to monitor and manage all developments has diminished accordingly. Indeed, it is the hallmark of a good executive to delegate responsibility, as long as those being delegated are capable of doing the job. With issues involving hundreds of millions of dollars merely a rounding error in American budgets, tertiary officials dealing with those so-called lesser issues can affect the fate of programs, and indeed countries, for whom $100 million is not small change. Such was the case with the Lavi, and such will be the case with other future programs, such as the Arrow missile.

My planned trip to Israel for the 1987 Sukkot holiday would have taken place within weeks of the cabinet vote. Friends and relatives strongly advised me not to expose my family to possible harm by angry demonstrators. I decided to postpone the trip. When we finally went the following year I received a few catcalls from passing dri-vers, but the vast majority of people I met were warm, friendly, and supportive.

One who was not was Moshe Arens. He could not come to terms with his defeat. He immediately resigned from the government in protest, stating that he could not be a partner to what he termed "this tragic decision," and that to remain in government would be "a betrayal of what I believe." He asserted that it was "nonsense" to say that the F-16 was as good as the Lavi, which he claimed "would have been the best warplane in the world." With no attempt to hide his bitterness, he added that "dreamers built this country and will continue to build it. . . . this wretched decision is a defeat of the dreamers by people of little faith."[2]

Shamir did not want to see him go and tried to talk him out of his decision. It was clear even then that Shamir would continue to keep the door open for him, and that Arens would one day return to the government. His chance to return came soon, and he was appointed foreign minister. When Likud obtained a stronger hold over the cabinet in the next elections, Rabin gave up the defense ministry and Arens replaced him. Six weeks after the Lavi vote, Arens had written in the *Jerusalem Post* that Israel's "fatal mistake" could "still be corrected."[3] For years he continued to insist that the Lavi should be revived, but the government would only support a program for developing Lavi components, which were tested on the prototype aircraft. Once Israel bought additional F-16s, the Lavi debate was finally closed.

Whenever I visited Israel in the years after the Lavi decision, one politician or another would tell me how Arens remained obsessed with his plane, and how much he resented my role in its cancellation. His Likud colleagues merely described him in unflattering terms; Labor leaders were not as nice about him.

I continued to visit the defense ministry during the years he was in charge there. But we never met. The Israeli press continued to ply me with questions about the Lavi. Did I still think the decision was a good one? (Yes.) Did I regret my role? (No—I was proud of it.) Was it good for Israel? (Yes.)

I also kept in touch with Yitzchak Rabin during his brief period in the political wilderness before he again formed a government as prime minister. He would delight in telling me the latest story about Arens's discomfiture. One day, as he recounted yet another story of Arens's hopes to revive the plane, I tried to respond in kind. "You know," I told him, "my father has known Arens for years; he knew Arens's father. So you see, it's not just that Arens considers me a traitor, he feels I'm a traitor to the family."

"I know," came the gruff reply.

I was stunned. I just stared at the man. Rabin didn't bat an eyelash, didn't smile.

"Arens told me," he deadpanned.

Many of the principal actors in the Lavi drama, particularly those in Israel, remain active on the political scene. Shimon Peres, prime minister from 1984 to 1986, again became prime minister, having succeeded Yitzchak Rabin when he was cut down by an assassin's bullet. Moshe Arens, who resigned from the government upon the cancellation of the Lavi in 1987, later returned as foreign minister and then as defense minister under Prime Minister Yitzchak Shamir. Arens retired from politics after the Likud defeat of 1992. David Ivri, who served as MoD director general under both Arens and Rabin, continues to hold that office. Yitzchak Shamir stepped down after six years as prime minister following Likud's 1992 defeat, but served as a member of the Knesset until 1996. Former Likud finance minister Yitzchak Modai and Moshe Nissim are currently still in the Knesset.

The Lavi that Arens almost created has not disappeared, nor has the friction that it engendered between Israel and the United States. Israel reportedly exported Lavi technology to South Africa and, more ominously, to China. The Tel Aviv–Beijing connection has been particularly galling to American officials, and has remained an open sore as American-Chinese relations have deteriorated in recent years.

In any event, Israel seems to have reaped some benefit from the Lavi, as we had predicted a decade ago. Israel Aircraft Industries has suffered from some contraction, but hardly to the degree that it predicted. The company remains at the forefront of many technological developments, notably in the realm of unmanned aerial vehicles, and has become an increasingly important supplier to the U.S. Department of Defense. At the same time IAI has diversified into the commercial sector, while other Israeli companies, once starving for contracts because of their focus on the Lavi, have seen their fortunes revive. Those that have not have been the state-owned firms that have been mired in inefficiency and are targeted for privatization. Most important, most analysts agree that had the Lavi not been canceled, Israel would have found itself in a severe budget crisis that would have forced it to cancel other critical defense programs.

A healthy defense budget remains vital to Israel's security. Of course, Israel now finds itself deeply enmeshed in a peace process that has brought forth the tangible fruit of a new relationship with the Palestinians, peace with Jordan, a dialogue with Syria, and an end to the secondary boycott that all its Arab neighbors had imposed on it since 1948. Nevertheless, even a comprehensive peace with the Arabs will not guarantee security indefinitely. Indeed, the price of peace is already driving up defense costs as Israel redeploys its forces from the West Bank. Israel's need to acquire antiballistic missiles, sophisticated satellites, and other cutting-edge high-technology systems will add further pressure to a defense budget that is under increasing competition from social programs spurred by new immigration from Ethiopia and the former Soviet states. Viewed in this context, the Lavi would only have been a budgetary albatross, weighing down Israel's ability to meet the new defense demands of the twenty-first century.

Israel has seen more than its share of war and misery during its brief existence. But it also has some very good friends around the world, new as well as old, many of whom demonstrated that friendship by attending the remarkable state funeral for Yitzchak Rabin on November 7, 1995. Nevertheless, none of these friends has been consistently more supportive of the Jewish state than the United States. It was Israel's good fortune that in 1987, America acted in character and as a best friend, and urged Israel to stop the Lavi. That Yitzchak Rabin did so, in the face of controversy, bitterness, and invective, was yet another hint of the man's greatness, which subsequently shone so brightly in the far more difficult, demanding, and complex context of Middle East peacemaking, until it was brutally snuffed out on that warm November night in 1995.

Notes

Chapter 1: ISRAEL'S NEW FIGHTER

1. Oded Eran, who served as the embassy's liaison to Congress, quoted in Charles R. Babcock, "How U.S. Came to Underwrite Israel's Lavi Fighter Project," *Washington Post*, 7 August 1986, A1.
2. The legislation was formally sponsored by Congressman Jack Kemp of New York and Senate majority whip Russell Long of Louisiana. See Peter Hellman, "The Fighter of the Future," *Discover*, July 1986, 46.
3. Babcock, "How U.S."
4. Ibid.
5. Some reports state that the meeting lasted seven hours; see ibid.
6. Hellman, "Fighter," 50.

Chapter 2: SOMETHING ABOUT THIS PROGRAM STINKS

1. As of mid-April 1985, Israel Aircraft Industries and its subsidiaries were supplying data entry electronic units, constant speed drives, and airframe components for the F-16. The other Israeli subcontractors were Israel Military Industries, Elbit, Orbit, TAT, and Cyclone Aviation Products. Coproduction systems that would be included in Peace Marble II as Israeli-government-furnished equipment were being produced by IAI and its subsidiaries, IMI, Tadiran, Astronautics, and other Israeli firms. Lt. Col. Ransom, "F-16 Major Contractors and Subcontractors," unpublished DSAA paper, April 15, 1985; unclassified.

259

Chapter 3: GETTING DOWN TO BUSINESS

1. Amir Oren, "Make War, Not Lavi," *Davar,* 26 April 1985.
2. Interview with Moshe Arens, Jerusalem Domestic Radio (in English), April 1985.
3. David A. Brown, "Israelis Stress Need for U.S. Aid to Complete Lavi Development," *Aviation Week & Space Technology,* 25 March 1985, 18–19.
4. Ibid., 20.
5. Eitan Haber et al., "Red Line Reached; Cancellation of 'Lavi' Possible," *Yediot Achronot,* 25 November 1984.

Chapter 4: MEETINGS, MEETINGS—IN WASHINGTON, IN ISRAEL

1. Martin A. Meth, "Memorandum for Dov Zakheim: LAVI R&M and Logistic Question," 18 June 1985; unclassified.
2. Harry Courtwright, "Letters: Lavi Fighter," *Aviation Week & Space Technology,* 10 June 1985, 138.
3. Letter from Levine to Weinberger, 9 July 1985.
4. Meth, "Memorandum . . . Lavi R&M."
5. Robert Waters, "P&W Share in Lavi Fighter May Rise," *Hartford Courant,* 2 July 1985.
6. Interview with Yitzchak Rabin, Kol Israel Radio (Jerusalem), 19 July 1985.
7. *The Babylonian Talmud, Makkoth* (London: Soncino, 1986), 24b.

Chapter 5: PREPARING THE LAVI REPORT

1. Letter from Meron to Iklé, 8 August 1985.
2. Letter from Iklé to Meron, 26 August 1985.
3. Letter from Meron to Iklé, 13 September 1985.
4. "Pentagon Officials Said to Review Israel's Lavi Program," *Aerospace Daily,* 22 August 1985, 290.
5. Letter from Gast to Ben-Joseph, 18 September 1985.
6. Cable from U.S. Defense Attaché Office, Tel Aviv, to SecDef;ASD//ISA; SecNav, "Transcript of MoD Rabin and Secretary Lehman's remarks to Press," 191212Z April 1985, cite 1042; unclassified.
7. Memorandum from the under secretary of defense, 16 December 1985; unclassified.
8. Proverbs 14:28.
9. "DoD Seriously Considering Israeli Lavi as New USAF Fighter, Gets FY-87 Funds," *Inside the Pentagon,* 18 October 1985, 1.
10. "Lavi Fighter in the Running for A-10 Replacement Contest," *Defense News,* 21 October 1985, 2.

11. "Israel Requests Transfer of U.S. Funds Designated for the Lavi—Even If the Fighter Is Not Manufactured," *Ma'ariv*, 9 December 1985. Halperin never openly opposed the Lavi; he was much too clever in that regard.

12. James A. Russell, "Officials Doubt Lavi Price Estimates," *Defense Week*, 18 November 1985, 1.

13. Letter from Ben-Joseph to Gast, November 27, 1985. Ben-Joseph indicated that the engineering change proposal amounted to $38 million in 1982 dollars, which was true enough, but the actual disbursement requested was the larger sum. Moreover, the actual Grumman proposal appeared to be significantly higher than Ben-Joseph's letter indicated. Total costs in 1982 dollars amounted to $49.6 million, given Grumman's request for adjustments for unforeseen matériel costs and its fixed fee of 10 percent for the engineering change. A U.S. Air Force cost analysis of the proposed change put the total at $58 million in 1985 dollars. Memorandum from Col. M. Foley, AF/RDQT, to Zakheim, 10 December 1985; unclassified.

14. Grumman Aerospace Corporation, "LAVI Wing and Vertical Tail, Proposal for Production 3 July 1985," p. 2; appended to ibid.

15. Quoted in "Moda'i: The Lavi Project Will Fall, It Doesn't Stand a Chance, It Will Be of No Advantage," *Ha'aretz*, 18 December 1985.

16. Joseph Alpher of the Tel Aviv University Jaffee Center for Strategic Studies, quoted in "Israeli Fighter Plane Being Readied," *Journal of Commerce*, 6 December 1985, 17B.

17. Quoted in "Menachem Einy, Director of the Lavi Project: The Lavi's Prototype Will Fly Within 10 Months," *Globes*, 3 December 1985.

18. "Rabin Prefers Reduction or Cancellation of the Lavi Project If the Defense Budget Sustains Another Cutback," *Ha'aretz*, 8 December 1985.

19. The Lavi's capabilities were published in many journals during the period 1985–87. A good summary appears in Gerald Green, "Israel's Lavi," *National Defense*, December 1986, 77–79.

20. Office of the Deputy Under Secretary of Defense (Planning and Resources), *The Lavi Program: An Assessment of Its Mission, Technical Content and Cost*, unclassified Executive Summary, February 1986, ES-1,2.

21. Ibid., 15; unclassified.

22. Ibid., ES-4.

23. Ibid., 23; unclassified.

Chapter 6: ISRAEL RESPONDS TO THE LAVI REPORT

1. "Israel May Scratch Its Plans to Build 'Lavi' Jet Fighter," *Washington Times*, 4 February 1986, D4.

2. "Israel Considers Slowdown in Lavi Program," *Aviation Week & Space Technology*, 10 February 1986, 32.

3. "Israelis Dispute DoD's Cost Analysis of Lavi Fighter," *Aerospace Daily*, 14 February 1986, 255.
4. Hirsh Goodman, "IDF Says U.S. Erred in Estimate of Lavi Costs," *Jerusalem Post*, 19 February 1986.
5. Letter from Zakheim to Tropp, 2 April 1986.
6. Ze'ev Schiff, "How Much Will the Lavi Really Cost?" *Ha'aretz*, 19 February 1986.
7. Letter from Eini to Zakheim, undated, attached to letter from Gal-Golkind to Zakheim, 25 March 1986.
8. Dov S. Zakheim, "Memorandum for the Comptroller of the Air Force," 26 March 1986; unclassified.
9. Zakheim to Eini, 28 March 1986.

Chapter 7: RABIN BLESSES A SECOND LAVI REPORT

1. Israel Landers, "C'est Lavi," *Davar*, 4 April 1986.
2. Zvi Tinor, "The Lavi Is a National American Project," *Al Hamishmar*, 1 April 1986.
3. Letter from Joan M. McCabe, GAO associate director, to Weinberger, 20 March 1986; unclassified.
4. Wolf Blitzer, "Israel told to get biggest bang for the buck," *Jewish Week*, 21 March 1986, 3.

Chapter 10: RUN-UP TO ANOTHER STUDY VISIT

1. In mid-April, a midlevel DSAA official had proposed that we inform the Israeli purchasing office in New York that we cap Lavi FMS funding for contracts at $550 million, to reflect Rabin's cost cap. The result would have been a slowdown in Lavi contracts. At the time, I opposed making the policy official, arguing that to do so was premature.
2. Oded Granot, "U.S. Puts a Freeze on Contracts and Deals for Lavi Production," *Ma'ariv*, 23 May 1986, 1.
3. The Israelis reportedly were considering reducing the electronic warfare segment of the Lavi fighter, even though that was the area that they developed independently and guarded most jealously from American eyes. "Lavi EW Cuts," *Aerospace Daily*, 27 April 1986, 154.
4. David Silverberg, "Israeli Lavi Company Executives to Meet with McDonnell Douglas," *Defense News*, 28 April 1986.
5. Wolf Blitzer, "We Have a Problem with the Lavi," *Jerusalem Post*, 30 May 1986.
6. Hirsh Goodman, "Pentagon, IAI Wrestle over the '46 Percent Gap,'" *Jerusalem Post*, 30 May 1986.
7. George P. Shultz, *Turmoil and Triumph: My Years as Secretary of State* (New York: Scribner's, 1993), 443.

8. Hirsh Goodman, "Doubts About Lavi Jet Fighter Grow," *Jerusalem Post*, 4 June 1986. In addition to being revealed in its entirety on Israeli television, reports of the letter's contents and its implications appeared in, among others, the London *Daily Telegraph*, the *Financial Times*, *The Times*, and *Washington Times*.

9. Meir Merhav, "Message from the U.S.: Don't Expect Us to Keep Boosting Aid," *Jerusalem Post*, 30 May 1986.

10. A small newspaper item indicated as much: "Zakheim–Rabin in Crucial Lavi Talks," *Jerusalem Post*, 2 June 1986.

11. The Carter Doctrine was enunciated by the president in the aftermath of the Soviet invasion of Afghanistan. It stated that the United States would go to war to protect our interests in the petroleum resources of the southern Gulf. About ten years later that was exactly what we did.

Chapter 11: THE DEBATE HEATS UP

1. Hirsh Goodman, "Doubts About Lavi Fighter Grow: Zakheim Meeting 'Dismal Failure,'" *Jerusalem Post*, 3 June 1986.

2. Alex Fishman, "Defense Minister Rabin: Israeli Cost Estimates Not 100 PCT Accurate," *Al Hamishmar*, 5 June 1986.

3. Leonard Silk, "Military Costs an Economic Issue," *New York Times*, 4 June 1986, D2. The story was cited in *Davar* the following day.

4. Reported in Magda Abu Fadil, "Israel's Lavi Blunder," *Middle East*, June 1986, 15.

5. General Amir Drori, quoted in Ze'ev Schiff, "The Lavi Between the IDF and the Pentagon," *Yisrael Shelanu*, 6 June 1986, 1.

6. Tony Banks, "Lavi Program May Fall Victim to Gramm-Rudman," *Defense News*, 2 June 1986, 1.

7. See, for example, M. K. Dan Tichon's remarks in "Americans Urge Israel to Scrap Lavi Fighter Project," *New York City Tribune*, 11 June 1986.

8. Banks, "Lavi Program," 1.

9. The interview appeared in the paper's June 12 edition.

10. "Rabin: U.S. Seeks 'Change of Direction' on Lavi," Jerusalem Domestic Service in Hebrew, 1050 GMT, 11 June 1986.

11. Staff, "Rabin: Lavi Project Will Go On," *Jerusalem Post*, 11 June 1986.

12. Ibid.

13. Staff, "The Lavi," *Jerusalem Post*, 11 June 1986.

14. Reuven Pedatzur, "Commentary," *Ha'aretz*, 11 June 1986.

15. Silk, "Military Costs," D2.

16. *Jerusalem Post* staff, "Rabin: Lavi Project."

17. "Decrease in U.S. Aid to Israel May Force Halt in Lavi Program," *Aviation Week & Space Technology*, 9 June 1986, 26.

18. Ibid.

Chapter 12: ENTER THE GAO

1. James A. Russell, "Dispute Still Raging over Israeli Lavi," *Defense Week*, 16 June 1986, 7.

Chapter 13: ROLLOUT

1. Hirsh Goodman, "Defense Ministry Checking Lavi Costs," *Jerusalem Post*, 17 June 1986.
2. Letter from Kasten to Iklé, 27 June 1986.
3. Goodman, "Defense Ministry."
4. Paul Bedard, "Lavi Under Fire," *Defense Week*, 21 July 1986, 11.
5. Ibid.
6. Ibid.
7. Dan Fisher, "Israel's Controversial Lavi Fighter Debuts," *Los Angeles Times*, 22 July 1986, Part IV, 1.
8. Hirsh Goodman, "Brass Bands, Doubts as Lavi Rolls Out Today," *Jerusalem Post*, 21 July 1991.
9. Moshe Shlonsky, Israeli Military TV, 3 August 1986.
10. Tali Zelinger, "The World's Most Expensive Jet Fighter," *Davar*, 8 August 1986.
11. Simcha Bahiri, "Displacing Milk and Honey," *Jerusalem Post International Edition*, 30 August 1986.
12. Charles R. Babcock, "How U.S. Came to Underwrite Israel's Lavi Project," *Washington Post*, 6 August 1986, A1.
13. Jerusalem Domestic Service in Hebrew, 1 August 1986.
14. Letter from Ben-Joseph to Zakheim, 14 August 1986.
15. Derek J. Vander Schaaf, deputy inspector general, to the Secretary of Defense, "GAO Review, 'Israel's Lavi Program,' GAO Code 464114—Information Memorandum," 25 August 1986; unclassified.

Chapter 14: THE NAVY STUDY IS COMPLETED

1. Cited in James A. Russell, "U.S. Tells Israel to Overhaul Its Navy Modernization; Rejects Aid Plea," *Navy News & Undersea Technology*, 21 November 1986, 1.
2. Yossef Valter, "Disagreements with the United States over Naval Modernization," *Ma'ariv*, 12 November 1986; Oded Granot, "Officials Grapple over Israeli Submarines; Funding in Limbo," *Defense News*, 17 November 1986, 3.
3. Valter, "Disagreements," 1.
4. Yossi Melman, "Shamir Supports Israeli Modernization Program," *Davar*, 11 November 1986.
5. "It's the Lavi or the Navy, Expert Says," *Jerusalem Post*, 13 November 1986, 1.

6. Ud Gundar, "The Lavi Engine Remains in Bet Shemesh as a Guarantee to Prevent the Plant's Closure," *Ha'aretz*, 13 November 1986.

Chapter 15: WE FINISH THE ALTERNATIVES STUDY

1. Ze'ev Schiff, "The Defense Industries in Trouble," *Ha'aretz*, 21 November 1986.
2. Brigadier General Robert Deligatti, USAF, quoted in James A. Russell, "'Kill The Lavi': Air Force General," *Defense Week*, 15 December 1986, 2.
3. Office of the Deputy Under Secretary of Defense, *The Lavi Aircraft: An Assessment of Alternative Programs*, unclassified Executive Summary, 21 December 1986, ES-1, 4.
4. "The Wait Goes On for Lavi Study," *Defense News*, 29 December 1986, 2.
5. James A. Russell, "Lavi Seen Breaching Israeli Cost Ceiling," *Defense Week*, 5 January 1987, 5.
6. Jerusalem Television Service, "Labor Costs Delaying Lavi Production," 22 December 1986.
7. Ron Ben Yishai, "The Lavi Makes a Preemptive Strike Against Its Main Opponent: Zakheim," *Yediot Achronot*, 2 January 1987.
8. Dan Fisher, "Test of Israel's Lavi Jet Fighter a Success," *Los Angeles Times*, 1 January 1987.
9. Hirsh Goodman, "Airborne, but a Rough Ride Ahead for the Lavi," *Jerusalem Post*, 2 January 1987. Within a week Goodman was citing projections that Israel would require only 150 Lavi aircraft (Hirsh Goodman, "Answers to Lavi Critics," *Jerusalem Post*, 9 January 1987, 5).
10. Goodman, "Airborne."
11. Ibid.; Reuven Pedatzur, "U.S. to Israel: Choose One of Five Alternatives to the Lavi," *Ha'aretz*, 2 January 1987.
12. *Davar* correspondents, "Zakheim Arrives Today—and Will Recommend a Half Dozen Alternatives to Lavi," *Davar*, 4 January 1987, 1.
13. Joseph Walter, Ron Dagony, and Don Arkin, "There Is No Substitute to the Lavi Alternatives; Cost of Project Cancellation: About 400 Million Dollars," *Ma'ariv*, 5 January 1987, 1.
14. *Al-Hamishmar*, 6 January 1987.
15. Akiva Eldar, Zohar Blumenkranz, and Reuven Pedatzur, "5,000 Will Be Saved at IAI If It Is Decided to Drop the Lavi in Favor of the F-16," *Ha'aretz*, 7 January 1987.
16. Arens, or one of his aides, seems to have leaked the story to *Yediot Achronot*, which strongly supported the Lavi. Roni Shaked, "Refuses to Meet with Arens—in East Jerusalem," *Yediot Achronot*, 8 January 1987.

17. Deputy Minister of Defense Zvi Ben Moshe, quoted in "1,700 and Not 3,000," *Davar*, 9 January 1987.
18. Ron Ben Yishai, "The Real Alternative to the 'Lavi,'" *Yediot Achronot*, 9 January 1987.
19. Reuven Pedatzur, "Analysis of Lavi Alternatives in Hands of Group Interested in Lavi Development," *Ha'aretz*, 8 January 1987.

Chapter 16: EXITING DOD

1. Hirsh Goodman, "Answers to Lavi Critics," *Jerusalem Post*, 9 January 1987.
2. Levy's changed position did not go unnoticed in the United States. See Mary Curtius, "U.S. Steps Up Efforts to Get Israel to Scrap Jet Project," *Christian Science Monitor*, 9 January 1987.
3. Reuven Pedatzur in *Ha'aretz*, 3 February 1987.
4. "Israel's Next Airplane," *Washington Post*, 19 January 1987, A18.
5. "Top Official Dismisses 'Fantastic' U.S. Ideas," *Jerusalem Post*, 15 January 1987.
6. Goodman, "Answers." Moshe Arens made some similar arguments in an interview in *Defense Week*, 19 January 1987, 24.
7. Richard Smull, "Talking Points: Israeli Lavi Fighter Aircraft," undated memorandum; unclassified.
8. "Lavi Passes Test," *Near East Report*, 12 January 1987, 7.
9. "The Lavi: More Than Just the Money," *Young Israel Viewpoint*, February 1987, 7.
10. Phil Baum and Raphael Danziger, *The Lavi Jet Fighter Controversy and Its Impact on U.S.–Israeli Relations* (New York: American Jewish Congress, 1987), 6.
11. Dan Margalit, "A Prime Minister on the Sidelines," *Ha'aretz*, 8 January 1987.
12. Committee on Foreign Affairs, "Hamilton Releases GAO Report on Cost Estimates for Israel's Lavi Aircraft," press release, 23 February 1987.
13. Reported in *Yisrael Shelanu*, 3 March 1987, 8.
14. Mendel Weinbach, "David's Slingshot," *Jewish Press*, 20 March 1987, 22A.
15. Jonathan Broder, "Army in Israel Eyes U.S. Jets," *Chicago Tribune*, 4 March 1987.
16. See, for example, James A. Russell, "Lavi May Be Canceled Soon," *Defense Week*, 9 March 1987, 12.
17. "Lavi Debate Boils Over," *Defense Week*, 9 March 1987, 2.

Chapter 17: THE LAVI IS GROUNDED

1. "Lavi Linked to South Africa Policy," *Ha'aretz*, 29 March 1987.
2. "The Lavi: Why Does Rabin Stall?" *Ha'aretz*, 25 March 1987.

3. Reuven Pedatzur, "American Jewish Organizations Oppose a Lavi Investment Fund," *Ha'aretz*, 25 March 1987.

4. Reuven Pedatzur, "Defense Establishment Enquiry: Production of 150 Lavi Is 50% More Expensive Than an F-16 Purchase," *Ha'aretz*, 22 April 1987.

5. James Bernstein, "Israeli Jet's Shaky Outlook Worries U.S. Defense Firms," *Newsday*, 26 May 1987.

6. *Defense News*, 8 June 1987.

7. "The Government Has Yet to Decide on the Lavi Matter," *Yisrael Shelanu*, 26 June 1987, 2.

8. Andrew Meisels, "Israeli Warplane to Cost Double," *Washington Times*, 1 July 1987.

9. Bruno is quoted in Dan Fisher, "Key Israel Vote Supports Fighter Plane," *Los Angeles Times*, 10 August 1987.

10. Joshua Brilliant, "The Lavi Debate Now a Search for Compromise," *Jerusalem Post International Edition*, 11 July 1987.

11. "Ministers Sharon and Arens Defended the 'Lavi' Before Members of the Foreign Affairs and Defense and Finance Committees," *Yisrael Shelanu*, 31 July 1987, 2.

12. Z. Sasson, "The Ministers of Defense and Finance Warn Against Continued Development of the Lavi," *Erev Shabbat*, 14 August 1987, 31.

13. State Department spokesman Charles Redman, quoted in Michael R. Gordon, "U.S. Is Pressing Israelis to Drop Costly Jet Effort," *New York Times*, 12 August 1987, A1.

14. Shultz, *Turmoil and Triumph*, 443.

15. "American Ultimatum to Israel to Cancel the 'Lavi' Jet Project," *Yisrael Shelanu*, 14 August 1987, 1.

16. Roni Shaked, "Ten For, Ten Against, Four Waffle," *Yediot Achronot*, 14 August 1987.

17. James Rupert, "Israelis Postpone Lavi Vote," *Washington Post*, 17 August 1987, A13.

18. Andrew Meisels, "Israelis Will Scrap Costly Warplane Project," *Washington Times*, 31 August 1987.

Chapter 18: AFTERMATH

1. Larry Cohler, "Post-Lavi Israel–U.S. Cooperation Falls Short of Hopes," *Washington Jewish Week*, 10 December 1987, 4.

2. Quoted in Andrew Meisels, "Lavi Backer Quits Cabinet in Anger over Scrapping Jet," *Washington Times*, 3 September 1987.

3. Moshe Arens, "The Wreck of the Lavi," *Jerusalem Post*, 10 October 1987.

Index

A-4 attack aircraft, 5, 82
A-7 attack aircraft, 27
A-10 attack aircraft, 27, 36, 75–76
Abe, Shintaro, 147
Abrahamson, James, 183
Ackerman, Gary, 169, 173
Ad-Dustour, 217
Aerospace Daily, 66, 89, 112
Aguda Party, 222
Agunah, 71
Air Combat Command, 24
Air Force (U.S.), 18, 26, 27, 75, 76, 103, 187, 211; Aeronautical Systems Division, 129, 131, 179, 181–83; Air Staff, 24, 126, 208; Systems Command, 129, 179, 203
Akiva, Rabbi, 64
al-Aqsa Mosque, 41
Aldridge, Edward (Pete), 129
Algeria, 61
Al Hamishmar, 102, 151, 154, 216
American Enterprise Institute (AEI), 53, 128, 204
American Israel Public Affairs Committee (AIPAC), 8, 9, 13, 28, 29, 30, 31, 77, 108, 116, 165, 216, 229
American Jewish Congress, 106–108 passim, 230
Antigua, 67

Anti-Defamation League, 157
Anti–tactical ballistic missile (ATBM), 183, 207, 214–215, 247–48, 254
Arad, Moshe, 250
Arbelli-Almozlino, Shoshanna, 251–52
Arens, Moshe, 21, 30, 37, 55, 77, 78, 89, 155, 162, 165, 174, 177, 179, 198, 217, 220–21, 228, 231, 257; ambassador to U.S., 8, 9, 230; and Congress, 13; and D. Levy, 243; and DZ's father, 118; and IAI, 7; and Iklé visit, 35–36, 46; and Lavi sales to U.S., 76; and Shamir, 14, 222, 256; and Shomron, 87, 147–48, 235; attends Lavi roll-out, 210; becomes minister without portfolio, 15; briefs Knesset committee, 248–49; chairs Knesset committee, 6; criticizes DZ, 172–73; expects more U.S. funds for Lavi, 102, 108, 244; leads Likud support for Lavi, 240, 241; meets DZ, 42; resigns from government, 255; succeeds Sharon, 10–12; tries to revive Lavi, 256
Armitage, Richard, 17–18, 20, 98, 128, 171–72, 226, 230, 232, 234–35, 238, 244, 247, 250, 254
Arms Control and Disarmament Agency, 74
Arrow missile, 207, 254–55

Aryeh fighter, 5, 6
Aspin, Les, 171
AV-8B (Harrier) attack aircraft, 97, 129,
    131, 182, 202, 205, 207
Aviation Week & Space Technology, 40,
    55, 86, 112, 158
Avidan, Eli, 117
Australia, 21
AWACS aircraft, 29

B-2 bomber, 131
Bank of Israel, 244
Bath Iron Works Corporation, 134, 139,
    190
Bavaria, Joe, 62, 242
Bechtel Corporation, 122
Begin, Menachem, 6, 11, 13, 45–46, 121
Belgium, 181
Ben Gurion, David, 44
Ben-Joseph, Abraham, 68, 79, 84, 100,
    127, 160, 175–80 passim, 183, 186, 193,
    204, 207, 213, 246
Ben Shoshan, Avraham, 71, 141, 174,
    194–95
Bernstein, Louis, 234
Betar movement, 11, 118
Bet Shemesh engine plant, 59, 60, 84, 90,
    177, 198, 208, 213–14
Bin Nun, Avihu, 43, 178, 226, 241
Blitzer, Wolf, 77, 106, 113, 139
Blumkine, Moshe (Bully), 40, 52, 87, 174
Bolles, Mike, 141
Bond, Jim, 9, 31, 165–66, 255
Boschwitz, Rudy, 25–26, 86
Brookings Institution, 29
Brown, Harold, 5, 8, 143
Bruno, Michael, 244–45
Brzezinski, Zbigniew, 121–22
Bunning, Jim, 167
Bush, George, 133, 189; administration
    of, 17, 24

Camp David Accords, 6, 122, 147
Canada, 4, 15, 192
Carlucci, Frank, 255
Carter, Jimmy, 17, 121, 144; administra-
    tion of, 29; Doctrine, 144
Central Synagogue (London), 75
Chagall, Marc, 222
China (People's Republic of), 147, 257
Chief Rabbinate (Israel), 71, 74
Chu, David, 24, 26
Clifford, Clark, 121

Clinton, Bill 125; administration of, 128
CNN (Cable News Network), 113
Columbia University, 122
Conference of Presidents of Major Jewish
    Organizations (Presidents' Conference),
    236, 240
Congress (Capitol Hill), 10, 19, 22, 26,
    28, 55, 111, 133, 165, 227, 239, 242,
    245, 246, 248; and Arens, 8–9, 13; and
    funding for Lavi, 25, 36, 107, 216; and
    GAO report, 159, 163, 184, 232; and
    Israeli naval modernization, 73; and
    Lavi contracts, 171, 72, 175; Lavi brief-
    ings for, 169–70, 228, 232–33; Rosenne
    and, 230. *See also* individual
    committees
Congressional Budget Office (CBO), 18,
    24, 30, 75, 99, 105, 109–111 passim,
    128, 143, 168, 184
Crossman, Richard, 88

Dachau concentration camp, 96
Dakar submarine, 71, 98
Davar, 34, 70, 101–102, 106, 198, 212, 224
Defense Acquisition Board, 110
Defense Guidance, 23–24, 51, 67
Defense News, 75, 112, 154, 172, 209,
    243
Defense Security Assistance Agency
    (DSAA), 20, 25, 68, 76, 100, 127, 151,
    177, 186
Defense Systems Acquisition Review
    Committee (DSARC), 110
Defense Week, 77, 78, 112, 173, 209
de Gaulle, Charles, 61
Denmark, 181
Department of Defense (Pentagon), 15,
    17, 20–24 passim, 29, 41, 50, 62, 72, 77,
    78, 90, 97, 101, 129, 132, 136–39 pas-
    sim, 149, 153–54, 158, 165–66, 175, 177,
    187, 204, 209, 216, 233–34, 239, 248–
    50 passim, 257; and ATBM, 183; and
    GAO study, 162–63; and Israeli naval
    modernization, 72, 97, 191, 193; and
    Lavi contracts, 151, 171–72; and Pol-
    lard, 124–25; DZ in, 109–113 passim;
    DZ leaves, 237–38; GAO and, 104–105,
    159; Ivri and, 102–103; Lavi report
    (1986), 82, 94; Office of Inspector Gen-
    eral, 184, 209; Office of International
    Security Affairs (ISA), 10, 91, 238;
    Office of Program Analysis and Evalu-
    ation (PA&E), 24, 26, 27, 52, 103;

Department of Defense (Pentagon),
(continued)
Office of the Secretary of Defense
(OSD), 13, 24, 26–28 passim, 48, 71,
75, 99, 109, 189; Peres visits, 189;
Shamir visits, 231; supports Lavi ter-
mination, 127
Department of Justice, 125
Department of State, 17, 28, 29, 48, 52,
78, 91, 116, 160, 171, 175, 186, 189, 209,
231–32, 235; and Pollard 125; Bureau
of Near East Affairs (NEA), 78, 135,
172; Bureau of Politico-Military
Affairs (P/M), 78, 135; calls for Lavi
cancellation, 249; GAO and, 104;
Shultz at, 122
Diaspora, 3; Museum of the, 198–99
Dine, Tom, 29–30, 77
Disney World, 140
Djibouti, 146
Dole, Robert, 17
Dolphin class submarine, 134, 189,
191–93
Dome of the Rock (see Mosque of Omar)
Dornan, Robert, 169
Dulles, John Foster, 121

Eban, Abba, 243, 248
Egypt, 47
Eini, Menachem, 37, 50, 58, 70, 87, 92–
93, 131, 149, 152, 174, 176–77, 186–88,
204, 209, 211–15, 222–23, 228, 242;
apologizes to DZ, 245, 247; clashes
with Hardy, 52–53; criticizes DZ, 227,
233–34; differs with IAI, 226; hosts
DZ, 89; receives first Lavi briefing, 38
Eisenhower, Dwight, administration of,
121
Eisenstat, Stuart, 121–22
Electric Boat shipyard, 59
El Salvador, 61, 140
England, 21
Entebbe, Israeli raid on, 32
Eritrea, 146
Ethiopia, 258
European Program Group (EPG), 181, 185,
206
European Union, 121

F-4 fighter, 82
F-5 fighter, 7, 131
F-5G fighter, 7

F-14 fighter, 27
F-15 fighter, 6, 26, 27, 36, 76, 97, 129,
131–32, 202–208 passim, 211
F-16 fighter, 6, 12, 13, 26–28 passim, 35–
36, 38, 50, 58, 67, 86, 97, 126–27, 129,
132, 157, 161, 169, 179–82 passim, 185,
202–207, 211–212, 215, 221, 224, 227–
28, 239–40, 246, 251, 254, 256; "Peace
Marble" program, 18, 19, 182, 202, 205,
207, 214, 235, 254; radar, 222
F-18 fighter, 26, 27, 28, 97, 129, 131, 182,
202–203, 206
F-20 fighter, 7, 99, 129, 131, 182, 203, 207
F-100 engine, 12
F-404 engine, 5
Falklands War (1982), 20
Federal Bureau of Investigation (FBI),
123–24, 239
Finchley Synagogue (London), 75
Fischer, Stanley, 128, 204
Foley, Mike, 26, 52, 59, 128, 129, 136
Foley, Tom, 26
Foreign Assistance Act (1986), 102
Forrestal, James, 121
Four Powers Armaments Directors, 95
France, 95, 113, 196
Fuerza Aerea, 157

Gabriel missile, 196
Gal class submarine, 71, 72, 73, 98, 140,
192
Galilee, 143
Gast, Philip, 20, 25, 68, 79, 84
Gehinnom, 63, 117
General Accounting Office, 104–105,
142, 159–63, 166, 184, 208–209, 212–
213, 223, 228, 231
General Dynamics Corporation (GD), 12,
18, 50, 51, 68, 126, 129, 131–32, 173,
179–82 passim, 185, 202–203, 206,
222, 228, 231–33, 240
General Electric Corporation, 222
Germany, 72, 73, 95, 140, 193, 254
Golan Heights, 142–43, 145
Goodman, Hirsh, 89, 91
Gonzalez, Felipe, 147
Goren, Shlomo, 74, 75
Gramm-Rudman-Hollings legislation,
83, 107, 137
Grand Hotel, 100, 245
Great Britain, 95, 197
Great Synagogue (Jerusalem), 74

Gripen fighter, 37, 59, 188
Grossinger's, 51
Grumman Corporation, 7, 8, 68, 79, 80, 132–33, 173, 187–88, 227, 242

Ha'aretz, 81, 156, 198, 203, 221, 224
Haber, Eitan, 43, 100
Hadashot, 215
Haganah, 11
Haig, Alexander, 122
Halacha, 74
Halperin, Danny, 10, 76, 165, 175, 230
Ham, Steve, 62
Hamilton, Lee, 104–105, 159, 228–29, 231–32
Hardy, Linda, 26, 27, 48, 54, 59, 92–93, 136, 208; clashes with Eini 52–53
Hart, Gary, 170
Hartford Courant, 59, 60
Hauser, Miki, 33–34, 199
Hauser, Neal, 33–34, 199
HDW AG, 134, 189, 192
Hefetz, Shimon, 100, 196–97, 213, 223
Herut Party, 46; in U.S., 229
Hertzberg, Arthur, 53–54
Heseltine, Michael, 247
Heston, Charlton, 141
Hicks, Donald, 99
Higgins, William (Rich), 207
Hillel, Rabbi, 191
Histadrut Labour Federation, 130, 153, 171
Hitler, Adolf, 219
Hitler Youth Air Corps, 96
Hod, Mordechai, 242
Holocaust, 96, 118, 198
House Appropriations Committee, 184; subcommittee on foreign operations, 105, 174, 228
House Foreign Affairs Committee, 25, 55, 231; subcommittee on Europe and the Near East, 104, 159, 169, 232
Huntington, Al, 105, 106, 159–63
Hurevitz, Yigal, 251
Hussein, Saddam, 18

Icenogle, Larry, 109, 112, 172
Iklé, Fred, 3, 15–16;20–22 passim, 28, 31, 48, 77, 78, 97, 99, 126, 157–58, 183, 210, 226, 228, 244–45; and Armitage, 17–18; and DZ, 4, 23, 69; and Kasten, 165–66; and Lehman, 74; briefs

Boschwitz (1985), 25–26; briefs
Boschwitz (1986), 86; correspondence with Meron, 65–66; DZ represents on DSARC, 110; leads trip to Israel, 32–36, 42, 70; meets with Arens, 42; meets with Rabin, 47
India, 16, 29, 61
Ingalls shipyard, 134, 254
Ingenieurkontor Lubeck (IKL)AG, 73
Inouye, Daniel, 31, 166, 229
Inside the Pentagon, 75, 112, 248
Iraq, 6, 18; Scud attacks on Israel, 207
Israel, 6, 9, 20, 28, 51, 55, 60, 65, 70, 77, 84, 90, 91, 94, 130–31, 154, 179, 195, 201, 209, 216–18, 231, 233, 237–39, 242–43, 245–49, 255, 257–58; and DZ private life, 116–18; and GAO, 106, 160–63, 184, 223, 232; and Lavi contracts, 171–72, 177; and Pollard, 125; and U.S., 48, 152, 153, 156, 211; Boschwitz and, 25–26; contractors trip to (1986), 186; Dine and, 30; DZ attitudes to, 3–4, 113; DZ's first visit to, 32; DZ leads team to (7/85), 61; DZ leads team to, (2/86), 86–89; DZ leads team to (5/86), 134, 136–37, 149, 164; DZ leads team to (11/86), 194; DZ leads team to (1/87), 207, 210–15 passim, 225; F-16 and, 18–19, 132, 205, 227–28, 256; Grumman and, 187; Hamilton and, 105; Hart and, 170; industrial base, 155, 175, 235; Inouye and, 31; Kasten and, 165–66; Levine and, 233; naval modernization, 73, 98, 140, 192–93, 254; Presidents' Conference and, 240; reaction to Lavi cancellation, 253; Shultz and, 150, 214; U.S. aid to, 83, 138, 244; U.S. funding for Lavi in, 8; Weinberger and, 121, 150, 234
Israel Air Force (IAF), 53, 54, 80, 81, 211, 235, 236–37
Israel Aircraft Industries (IAI) 35, 45, 55, 60, 80, 84, 87, 93, 117, 135, 137–40 passim, 154, 170, 188, 199, 212, 216, 222, 235–36, 239–40, 251, 254, 257, and GAO, 160–63 passim; and Pratt & Whitney, 90; and termination costs, 228, 245; Arens and, 6–7; auditor general criticizes, 237; difficulties with Lavi integration, 131; DZ visits, 40–41; Eini and, 52, 152, 187, 226; estimates

Israel Aircraft Industries *(continued)*
job losses, 221, 224; founded, 5; GD
and, 12; Grumman and, 79–80, 227,
242; in Lavi alternatives study, 205;
Ivri and, 11, 103; Klemow and, 9–10;
Lavi rollout, 133–34, 173–74; Lavi
team, 50, 69–70; markets Lavi, 157–58,
243; Nissim and, 241; U.S. contractor,
18–19, 102; wage rate increases, 203–
204, 209–211; workers unions, 217,
249, 253
Israel Defense Forces (IDF), 11, 36, 43,
71, 86, 142, 143, 152, 178, 215, 218,
226, 235, 241, 249
Israel Navy (IN), 75, 84, 92, 140, 189,
193–95, 197, 210
Israel Shipyards Ltd.(Haifa shipyard), 72,
73, 139, 140–41, 190
Israeli-Lebanese Agreement (1983), 122
Ivri, David, 10–11, 40, 43, 70, 86, 87,
102–104, 106, 132, 142, 176, 194, 196,
214–15, 242, 254, 257

Jakobovits, Chief Rabbi Lord Immanuel,
21
Japan, 189
Jerusalem Post, 77, 89, 107, 113, 134, 136,
138, 150, 155–56, 170, 225, 247, 256
Jerusalem Television Service, 209
Jewish New Year, 191
Jewish Theological Seminary, 122
Jewish Week (New York), 113
Jews, 113, 122–23
Joint Chiefs of Staff, 20, 71, 144; Joint
Staff, 13, 72, 73, 98
Joint Political Military Group, 104
Joint Security Assistance Planning
Group, 104
Jordan, 258

Kasten, Robert, 9, 31, 165–66, 169,
175–76
Kelley, Paul (P. X.), 143–44
Kemp, Jack, 166, 169–70, 173–74, 229
Kennedy School of Government (Harvard
University), 90
Keret, Moshe, 131, 172–74, 177, 187,
241–42
Kerry, John, 153
Kfir fighter, 5, 6, 35, 57, 80, 157
Kissinger, Henry, 150
Klemow, Marvin, 9, 10, 13
Kloske, Dennis, 238

Knesset, 6, 41, 55, 135, 154, 195, 199, 211,
217–18, 221–22, 239–41, 243, 248–49,
257
Koch, Ed, 168
Kol Israel radio, 212

Labor Party, 14, 44, 45, 101, 135, 138, 152,
153, 195, 217, 239, 243, 251–53, 256
Lamm, Norman, 234
Lapidot, Amos, 40, 81, 152, 179, 210, 226,
237
Laromme Hotel, 62–63, 74
Latin America, 157
Lavi fighter, 15–17, 19, 22–23, 47, 56, 57,
66–67, 91, 128, 140, 189, 199, 234, 238–
40, 247, 254–55, 257–58; Alternatives
Steering Group, 126, 185, 188, 203;
alternatives to, 68, 157, 172, 180–83,
202, 205–208, 213, 216; and DZ private life, 109, 112–17, 119–20; and DZ
trip (5/86), 134; and GAO, 104–105,
160–63, 184, 223, 232, 233; and Iklé
visit, 35–36; and Israeli naval program,
70–73, 193–95, 197–98; and Presidents'
Conference, 236; and Rabin visit to
U.S.(11/85), 79; Arens and, 8–9, 76,
102, 108, 210, 223, 240, 241, 244, 256;
Armitage and, 18; as A-10 replacement
75, 76; auditor general's report on,
236–37, 244; Bet Shemesh engine and,
60; Boschwitz and, 26; Congress and,
25, 77, 81, 165, 215–16, 232; contracts,
92, 142, 151, 171, 175–77; critics of,
152, 153, 178, 217, 224, 227; Dine and,
30; DoD report (1986), 82–83, 192, 193;
DZ and Rabin discuss(5/86), 101; DZ
attitude to, 85; DZ briefing on, 37–38;
Eini and, 52; F-16 as competitor to,
126, 132, 179, 211, 221, 228; first flight,
129, 210; first flight delayed, 186, 188,
197, 203–204; flyaway costs, 86; foreign sales of, 158; Gast and, 20; GD
and, 173; Germany and, 97; grounded,
252–53; Grumman and, 133, 187; Hart
and, 170; Israeli industry opposes,
242–43; Israeli press and, 155–56, 215;
Jaffee Center report on, 137–38; Kasten and, 31, 166, 169; Kemp and, 229;
Lapidot and, 40, 226; "Lavi 2000,"
251; "lobby," 212; M. Levy opposes,
235; meaning of name, 3, 6; Neubach
and, 42; Nissim and, 220, 241;
Northrop and, 7, 55, 131; OSD/Air

Force cost estimates, 27–28; Peres and, 21, 45, 218, 248; production unit cost, 93–94; Rabin and alternatives to, 154; Rabin cost cap on 88, 100, 130, 139, 142, 161, 180, 204; Rabin favors, 21, 60–61; Rabin opposes, 249; religious community opposes, 220; rollout, 164, 170, 172, 184; Shamir and, 196, 250; Shomron opposes, 86–87; Shultz and, 135–36, 149–50, 214, 250; Steering Group, 48; supporters attack DoD study, 89–90; systems integration, 50, 59; Tamir and, 219; U.S. funds for, 78, 84; Weinberger and Rabin discuss, 99–100, 246; Weinberger opposes, 7, 127–28, 231; work breakdown structure (WBS), 50; Yaron opposes, 192

Lear Siegler Corporation, 7, 37–38, 59, 169, 186, 188

Lebanon, 45

Lebanon War (1982), 3, 13, 21, 178

Lee, Vernon, 132, 179, 185, 203

Lehi (Stern Gang), 11

Lehman, John, 7, 98, 167–68; and Israeli Navy program, 72–74, 139, 193

Lévesque, René, 16

Levine, Mel, 55, 169–71, 173, 233, 250

Levine, Robert, 128–29, 204

Levy, David, 231, 243

Levy, Moshe, 143, 145, 157, 178, 196, 210, 218, 226–27, 235

Lewis, Sam, 22, 31, 61, 66

Liberal Party (Israel), 45, 220

Libya, 72, 98

Likud Party, 6, 13, 14, 44, 118, 152, 153, 217, 220, 229, 231, 239–41, 243, 249–52 passim, 256–57

Limor, 141

Lithuania, 11, 118, 219

Litton Corporation, 134, 140

Loh, Mike, 24, 26

Los Angeles Times, 174

Luftwaffe, 95

Luttwak, Edward, 173

Ma'ariv, 114, 137, 156, 212–213

McDonnell Douglas Corporation, 50, 51, 68, 131–32, 182, 202–206 passim, 222, 228

McNamara, Robert, 90

Major, John, 247

Marine Corps (U.S.), 139, 143, 145

Marshall, George, 121

Massachusetts Institute of Technology (MIT), 6, 11, 110, 128, 204

Mastiff remotely piloted vehicle, 145

Meah Shearim, 219

Meth, Marty, 54

Merkava tank, 9, 84

Meron, Menachem (Mendy), 70, 87, 102, 213; correspondence with Iklé, 65–66

Merrill, Philip, 36, 46, 66

MiG-23 fighter, 249

MiG-29 fighter, 249

Ministry of Defense, Israel (MoD/Hakirya), 66, 92, 93, 106, 128, 142, 157, 176, 181, 183, 197, 216, 217, 234, 236, 240, 242, 244, 247; and GAO, 162–63, 184; and GD, 173; and Israeli naval modernization, 191, 193–96; and Lavi foreign sales, 131; contractor meetings with, 186, 222; DZ meetings at (6/86), 142; DZ meetings at (1/87), 214–15, 254; estimates Lavi cancellation's impact on employment, 221; Iklé team briefs, 37; Ivri and, 10, 102–104; Lavi program office 60, 69–70, 150, 164; meetings at Wright-Patterson AFB, 181; reports on Lavi wage rates, 203

Mirage fighter, 5, 58

Mirer Yeshiva, 219

Modai, Yitzchak, 42, 70, 88–89, 136, 164, 204, 220, 251, 257; DZ briefs, 45–46, 78, 81

Montefiore, Sir Moses, 63

Moog Corporation, 7, 174

Mosque of Omar (Dome of the Rock), 41

Mossad, 11, 147

Mulroney, Brian, 15

Nakasone, Yasuhiro, 147

National Security Council (U.S.), 48, 91, 171, 211

National Unity Government (Israel), 14, 44, 45, 152, 195–96, 231

NATO (North Atlantic Treaty Organization), 70, 168

Navy (U.S.), 26, 27, 71–73 passim, 139; OpNav (U.S. Navy staff), 73

Nazis, 11, 96, 118

Netherlands, 72, 181

Neubach, Amnon, 42, 43, 45, 70, 137, 185, 219, 245

New York Times, 59, 152, 156–57

Newsday, 242

Nissim, Yitzchak, 220–21

Nissim, Moshe, 204, 220, 240–45, 257 passim, 248–252 passim
Nobel Peace Prize, 101
North Warning System, 4, 15
Northrop Corporation, 7, 35, 50, 51, 55, 68, 99, 131, 173, 182, 203, 206–207
Norway, 181
Novick, Nimrod, 217

Obey, David, 105, 228–29
Office of Management and Budget (OMB), 42, 48, 52, 91, 107, 211
Offshore procurement funds (OSP), 13, 102
Operation Desert Shield, 18
Operation Desert Storm (Persian Gulf War), 18, 76, 207, 254
Oren, Avraham, 194–95
Orthodox Jews, 30, 41, 74, 115, 117, 119, 234
Orthodox Judaism, 29, 74
Osirak reactor (Iraq), 6, 18
Oxford University, 21, 33, 88

Pakistan, 16, 47
Palestine Liberation Organization (PLO), 147, 153
Papandreou, Andreas, 147
Passover, 22–23, 67, 99, 191
Pazner, Avi, 196
Peace Corps, 29
Pedatzur, Reuven, 156, 224
Pelletreau, Robert, 232
Peres, Shimon, 43, 135, 136, 155, 185, 215, 219, 231, 245; and Israeli cost estimates of Lavi, 151–52; and Rabin, 44, 185; and Shamir, 44–45, 195; and Shultz, 28, 135, 149; at Lavi rollout, 174; becomes prime minister (1984), 21; becomes prime minister, (1995) 257; DZ briefs on Lavi alternatives (1/87), 217–18; foreign minister, 204; on cabinet's Lavi task force, 242; "Lavi 2000" compromise, 251–52; seeks Lavi compromise, 247–48, 250; supports Lavi, 45, 136; visits Washington (3/86), 97; visits Washington (9/86), 189
Peretz, Yitzchak, 251
Perle, Richard, 124
Pickering, Thomas, 62, 70, 87, 88, 142, 149, 151, 156, 170, 172, 176–77, 197, 199–201, 212–214, 217, 223, 245, 251

Plains of Abraham, 16, 143
Poindexter, John, 171
Pollard, Jonathan, 114, 123–25, 145, 156, 230–31, 238–39
Powell, Colin, 79
Pratt & Whitney Corporation, 59, 90, 130
Pravda, 168–69
Purim, 119
PW 1120 engine, 59, 130

Quebec, 16, 143

Rabbinical Council of America, 234
Rabin, Yitzchak, 42, 46, 48, 53, 86, 97–98, 108, 128, 150, 152, 157, 164, 187, 203–204, 216, 226, 236, 243, 256–58; admonishes Eini, 234, 245; and alternatives study, 101, 126, 179, 197; and ATBM, 183, 214, 254; and Haber, 43; and Israeli naval program, 72–73, 196; and Kemp, 174; and Lavi contracts, 151, 170–72, 175–77; and Lavi termination liability costs, 181; and Peres, 44, 185; and South Africa, 239; and Weinberger, 99–100, 135, 142, 149–150, 174, 183, 213, 218; and Yaron, 192; assassinated, 257; becomes defense minister (1984), 14; caps Lavi cost, 88, 100, 130, 139, 142, 161, 180, 204, 210, 214; demands more funds for Lavi, 240–41; DZ briefs, (2/86), 88; favors Lavi, 21, 60–61; meets with DZ, (10/85), 70–71; meets with DZ (6/86), 141–42; meets with DZ (1/87), 213–15, 221–22; meets with Iklé, 47; misses Lavi first flight, 210; on cabinet Lavi task force, 242; opposes defense cuts, 81–82; opposes Lavi, 35, 228, 230–31, 233, 235, 250–51; reaffirms cabinet support, 164; rejects Jaffee Center report, 138; signs MOU with Carlucci, 255; speaks to Knesset, 154–56, 249; visits U.S. (11/85), 78–79; visits U.S., (6/87) 244–47
Rafael, 243
Ramadan, 191
Ram, Micha, 141, 195
Ramon, Haim, 153
RAND Corporation, 90, 204, 229
Rapid Deployment Force (RDF), 143–44

Rathenau, Walter, 123
Ratz Party, 135
Reagan, Ronald, 10, 13, 15, 21, 25, 28, 165; administration, 7, 24
Redd, John Scott, 36, 46–47
Reform Judaism, 29
Republican Party, 165
Roche, James, 173
Rockwell International Corporation, 134, 140
Rosen, David, 88
Rosen, Steve, 229
Rosenne, Meir, 118, 230
Rozhenoi, 11, 198
Rubenstein, Elyakim (Eli), 53, 118, 171, 196
Rubenstein, Miriam, 53
Rudd, Glenn, 25, 76
Russia, 61

Saar class corvette, 98, 134, 141, 191–93, 254
Sabbath (Shabbat, Shabbos), 23, 33–34, 62, 116–17
Sabra and Shatila (refugee camps), 8, 191–92
Sadat, Anwar, 147
St. Patrick's Day, 15, 16
Satmar Chassidim, 119
Sarid, Yossi, 135
Saudi Arabia, 29
Schapira, Avraham, 222
Schiff, Ze'ev, 91, 203, 209
Schnell, Karl, 95–97
Sella, Colonel, 239
Senate Appropriations Committee, 255; Foreign Operations Subcommittee, 31, 165, 166, 229
Senate Armed Services Committee, 170
Senate Budget Committee, 29
Senate Foreign Relations Committee, 153
Sephardim, 74
Shaked, Chaim, 191, 195, 252
Shamir, Yitzchak, 13, 21, 53, 221, 243, 245, 251, 257; and Arens, 14, 222, 256; and Zvi Zakheim (DZ's father), 233; and Israeli naval modernization, 196; and Lavi aftermath, 253–54; and Peres, 44–45, 195, 247–48; and Shultz, 213, 249–50; hosts DZ at home, 145–48, 149; replaces Begin, 11; visits Washing-

ton (10/85), 69; visits Washington (2/87), 231
"Shamrock Summit," 15, 16
Sharon, Ariel, 6, 7, 8, 10, 12, 13, 129, 231, 243, 248–49
Sharon, Emmanuel, 220
Shas Party, 251
Shitrit, Meir, 249
Shomron, Dan, 35, 86, 87, 178, 226, 235, 242
Shultz, George, 10, 28, 29, 78, 122, 152, 172, 213, 227, 244; and Lavi, 135–36, 149–50, 214, 250; and Shamir, 213, 249–50; and Weinberger, 29, 151, 170
Siegel, Seymour, 122
Singer, Israel, 236
Six Day War (June 1967), 4, 6, 32
Smith, Larry, 169–70, 173, 232, 250
Smull, Richard, 24, 52, 92, 129, 160, 204, 207, 211, 222, 224, 227, 236
South Africa, Republic of, 21, 147, 239, 257
Stalin, Joseph, 219
Stark, James, 171, 211
Stein, Herbert, 128, 204
Stern Gang. *See* Lehi
Strategic Defense Initiative (SDI), 183
Sukkot (Tabernacles), 67, 247, 255
Sweden, 4
Syria, 258
System Planning Corporation, 237

Tadiran Ltd., 242–43
Taft, William, 207, 238
Talmud, 23, 64
Tamir, Avraham (Abrasha), 218–19
Technion, 6
Tel Aviv University, 178; Jaffee Center for Strategic Studies, 137–38
Tel Nof Air Base, 239
Temple Mount, 41, 47, 143
Thatcher, Margaret, 147, 247
The Ten Commandments, 141
Thyssen AG, 134, 189, 192
Tisha b'Av, 56
Todd Shipyard, 134, 139
Torricelli, Robert, 169–70, 173, 233
Treblinka, 233
Tropp, Zvi, 52, 70, 87, 91, 107, 127–28, 150, 179, 234, 242
Trost, Carl, 167
Truman, Harry, administration of, 121

Tsur, Yaakov, 251
Turkey, 147

Union of Soviet Socialist Republics
    (Soviet Union/USSR), 5, 119, 219
United Nations, 61
United States of America, 4, 5, 8, 20, 21,
    58, 60, 74, 84, 91, 95, 132, 145, 172,
    187, 189, 211, 240; aid to Israel, 138;
    and F-15 sales to Israel, 6; and F-16
    sales to Israel, 6, 19, 240; and Israel,
    83, 113–14, 152, 153, 156, 171, 197,
    218–19, 227–28, 254, 257; and Israeli
    naval modernization, 72–73, 98, 139–
    40, 192; and Lavi program, 18, 30, 35,
    61, 126, 149, 157, 175, 184, 185, 211,
    244, 246, 249, 258; and Shamir, 146;
    embassy in Tel Aviv, 204, 212–13, 235;
    Jewish community in, 29, 125; Lavi
    proponents in, 22, 94; Lavi sales to, 76,
    131, 162; Lavi work in, 38
United Technologies Corporation, 59

Vander Schaaf, Derek, 184, 208
Vanunu, Michael, 197

Wall Street Journal, 59
Washington Institute for Near East
    Policy, 247
Washington Jewish Week, 113
Washington Post, 59, 178, 227
Washington Times, 59
Wehrmacht, 96
Weinberger, Caspar (SecDef), 4, 49, 55,
    60, 62, 78, 97, 116, 122, 129, 133, 184,
    187, 193, 226, 244; and alternatives
    study, 207–208; and Armitage, 17; and
    DZ, 123, 128, 210, 234, 247–48; and
    Israeli naval modernization, 73, 193;
    and Israeli public opinion, 91, 114; and
    Pollard, 125, 231, 239; and Rabin, 79,
    99–100, 135, 142, 149–50, 174, 213, 218,
    225, 245–46, 250, 255; and Shultz, 29,
    151, 170; Lavi contracts, 171–72, 177;
    opposes Lavi, 127, 231, 235; opposes
    Lavi export licenses, 7, 10
Weitzmann, Ezer, 5, 6, 78, 86–87, 136,
    164, 174, 221, 248–49
Weitzmann Institute of Science, 33,
    199–200

Western (Wailing) Wall, 47, 63, 88, 253
Westinghouse Corporation, 222
Wilson, Charles, 8, 9, 13, 166, 169, 173,
    250
Wolf, Charles, 204
Wolfe, James, 143
Wolfson, Sir Isaac, 74–75
Worcester Cathedral, 95
World Jewish Congress, 236
World War II, 11, 95
Wright-Patterson Air Force Base, 129,
    179–81, 183

Yaacobi, Gad, 242
Yaron, Amos, 191–92, 195, 230, 234–35,
    245–48
Yediot Achronot, 212, 224, 239, 250
Yemin Moshe, 62
Yeshiva University, 114, 234
Yisrael Shelanu, 152, 240
Yom Kippur, 67
Yom Kippur War (1973), 140
Yosef, Ovadia, 74
Young Israel, 229

Zakheim, Chaim, 118–119, 219, 225
Zakheim, Dov, 138, 150, 198–99, 209,
    211, 225; and Armitage, 17–18; and
    Dine 29–30; and Eini, 89, 227, 233–34,
    245, 247; and Gast, 20; and Iklé, 4, 23,
    69; and Kasten, 165; and North Warn-
    ing System, 15–16; and Rabin, 245–47;
    and Shamir, 145–48, 149; and Wein-
    berger, 123, 127; as Orthodox Jew, 114;
    at Oxford, 21; avoids Middle East
    issues, 3; contrasted with Pollard, 124;
    first visit to Israel, 32, 34; Lavi and pri-
    vate life of, 109, 112–17, 119–20; pro-
    moted, 69; "Zakheim report," 89
Zakheim, Joshua, 32, 119
Zakheim, Reuven, 116
Zakheim, Saadya, 117, 167
Zakheim, Zvi (author's father), 11, 12,
    118, 168, 219, 229, 236; and Shamir,
    233
Zakin, Dov, 199
Zionism 3,
Zucker, Dedi, 241
Zuza, Leonard, 107, 211

About the Author

Until March 1987, Dov S. Zakheim was U.S. deputy under secretary of defense for planning and resources. In that capacity, he played an active role in the Defense Department's system acquisition and strategic planning processes, and led the U.S. government's re-evaluation of Israel's Lavi fighter-bomber.

Currently, Dr. Zakheim is corporate vice president of System Planning Corporation (SPC), a high-technology research, analysis, and manufacturing firm based in Arlington, Virginia. He is also chief executive officer of SPC International Corporation, the consulting, international sales, and analysis subsidiary of SPC. In addition, Dr. Zakheim serves as consultant to the secretary of defense and the under secretary of defense for policy, and writes, lectures, and provides radio and television commentary domestically and internationally on American defense issues. He is an adjunct professor of international and public affairs at Columbia University and of political science at Yeshiva University. He is also an adjunct scholar of the Heritage Foundation, a senior associate of the Center for Strategic and International Studies, and serves on the advisory boards of the Nixon Center for Peace and Freedom, the Initiative for Peace and Cooperation in the Middle East, and the American Jewish Committee. He lives in Silver Spring, Maryland.